Separations Division 2013

Core Programming Topic at the 2013 AIChE Spring Meeting & 9th Global Congress on Process Safety

San Antonio, Texas, USA
28 April – 2 May 2013

ISBN: 978-1-62748-457-2

Printed from e-media with permission by:

Curran Associates, Inc.
57 Morehouse Lane
Red Hook, NY 12571

Some format issues inherent in the e-media version may also appear in this print version.

Printed by Curran Associates, Inc. (2013)

For permission requests, please contact AIChE
at the address below.

AIChE
3 Park Avenue
New York, NY 10016-5991

Phone: (203) 702-7660
Fax: (203) 775-5177

www.aiche.org

Additional copies of this publication are available from:

Curran Associates, Inc.
57 Morehouse Lane
Red Hook, NY 12571 USA
Phone: 845-758-0400
Fax: 845-758-2634
Email: curran@proceedings.com
Web: www.proceedings.com

TABLE OF CONTENTS

(33a) Pilot Plant Controlling Regimes and Scale-up Similitude

Jonathan H. Worstell, Consultant, Richmond, TX

We seldom determine the operating regime that controls a pilot plant. Thus, we may choose an inappropriate similarity when scaling to and designing a commercial-sized plant with pilot plant data.

This presentation discusses the need to determine the operating regime controlling a pilot plant and it shows how that controlling regime dictates the similarities used when designing a commercial-sized plant.

(33b) Degumming Process Design for Biodiesel Plants

Maoqi Feng, Div. of Chemistry and Chemical Engineering, Southwest Research Institute, San Antonio, TX

Phosphorus is a contaminant in biodiesel. Phosphorus originates from phospholipids in biodiesel feedstock. Phospholipids removal is required in biodiesel production. Degumming is the most commonly used method for phospholipids removal. Different degumming processes, such as acid hydrolysis, enzymatic degumming, membrane, ultrafiltration, and supercritical CO_2 processing, etc., can be selected for different feedstocks.

Continuous ultrasonic degumming improves mass transfer and thus higher efficiency. In-situ FT-IR analysis of phospholipids content provides a method for monitoring the phospholipids during degumming. Adsorbent could be used as a post-treatment method for quality assurance. Case study for scaling up the degumming process of soybean oil will be presented.

(33c) The Role of a Maxi Lab On the Scale-up of New Processes

César González Serrano, Matthew Kuhlman and Gregory Hemmer, SABIC, Mt Vernon, IN

The role of a Maxi Lab on the scale-up of new processes

The SABIC Innovative Plastic's High Performance Polymer Business constantly strives to engineer and commercialize new products for the marketplace, which often require that new processes are developed or existing ones are modified.

Initial laboratory trials leave questions opened to be addressed during pilot plant trials. However, availability, cost and feasibility may challenge the answers that can be realistically obtained from the pilot plant. A Maxi Lab provides an intermediate scale between the lab and the pilot plant where solutions can be found in a quick, visual and cost effective manner.

The utilization of the Maxi Lab allows for the early identification of critical steps in the operating procedures and preliminary scale-up parameters. Furthermore, due to the relatively small size of equipment, it allows for higher flexibility in changes of units of operation and multiple configurations of a piece of equipment. Such characteristics will contribute to reduce the number of runs required in the pilot plant and increase their effectiveness.

The Maxi Lab scale allows for production of a few kilograms of material, which grants collaboration with costumers via sample testing. These early samples are a key factor to understand the costumer CTQ's and unveil possible hidden CTQ's that may hinder the process at later stages.

(33d) Challenges in Commercial Scale up of Novel Poly(phenylene ether)-Polysiloxane Block Copolymer Process

Alvaro Carrillo and Scott Fisher, Noryl* Technology, SABIC Innovative Plastics, Selkirk, NY

Poly(phenylene ether) is an engineering thermoplastic known for its excellent hydrolytic stability, dimensional stability and inherent flame retardancy. A block copolymer of poly(phenylene ether)-polysiloxane (PPE-Si), produced by the oxidative coupling polymerization of 2,6 xylenol and polysiloxane, has been developed to deliver significant improvements in mechanical properties and flame retardancy.

Process development for new technology to culminate in production in an existing operating commercial chemical plant requires conscientious planning and risk mitigation efforts. This presentation illustrates the challenges involved in the commercial scale up of PPE-Si copolymer, covering areas such as: lab and pilot plant scale testing, determination of feasibility and process conditions for all unit operations, cross contamination of existing processes, management of transitions, quality specifications and testing methods, EHS and product stewardship considerations.

(45a) Pilot and Full-Scale Application of Technologies for Mercury Control

Noah Meeks, Research & Environmental Affairs, Southern Company Services, Inc., Birmingham, AL

With the final MATS rule going into effect in 2015, electric utilities must now consider mercury control as a needed technology in their emissions train. This talk wiill cover recent full-scale injections of activated carbon and alkali injection, as well as pilot-scale studies of non-halogenated carbons or non-carbon sorbents. We will also discuss recent mixed matrix membrane pilot studies. **(Most recent data will be included in the abstract closer to the meeting)

(45b) Acid Tunneling Vs Acidizing : A Case Study of Rudatain Oilfield-Kuwait

Adel Elsharkawy, Petroleum Engineering, Kuwait University, Kuwait, Kuwait and Mohamed Al-Hadher, Pet Eng, Kuwait Oil Company, Safat, Kuwait

Acid Tunneling Vs Acidizing : A Case Study of Rudatain Oilfield-Kuwait

By Mohammed Al-Hadher and Adel Elsharkawy

Kuwait University, College of Engineering & Petroleum

Po Box 5969 Safat 13060 Kuwait

Abstract

Acidizing is one of the most important techniques used in oil industry in order to enhance the reservoirs and increase oil production. There are three types of acidizing techniques which are matrix acid stimulation, acid fracturing, and acid tunneling. In this paper a comparison these types of acidizing is discussed. The case study of RA-0007 acid tunneling completed in Kuwait Oil Company is presented. It represents the first dendritic with extensive worm holes created by acid tunneling for maximum reservoir contact. The whole effect is aptly demonstrated by the change in PI of the well before and after the acid tunneling as mentioned before. Analysis of the well designed and implemented job showed that two tunnels were created successfully in MA-C formation through a two days rig-less operation. The results indicated an improvement in the well productivity index (PI) which had been increased from 1.3 to 3.87 bpd/Psi.

(45e) Sustained Casing Pressure Modeling and Well Integrity

Tony Rocha-Valadez, Mary Kay O'Connor Process Safety Center, Artie McFerrin Department of Chemical Engineering, Texas A&M University, College Station, TX, A. Rashid Hasan, Harold Vance Department of Petroleum Engineering, Texas A&M University, College Station, TX, C. Shah Kabir, EPT, Hess Corporation, Houston, TX and Sam M. Mannan, Artie McFerrin Department of Chemical Engineering, Mary Kay O'Connor Process Safety Center, Texas A&M University, College Station, TX

SUSTAINED CASING PRESSURE MODELING AND WELL INTEGRITY

T. Rocha-Valadez[1*], A.R. Hasan[2], C.S. Kabir [3] and S.M. Mannan[1]

[1]Mary Kay O'Connor Process Safety Center, Artie McFerrin Department of Chemical Engineering, Texas A&M University, College Station, Texas 77843-3122, US., email: *tonyrochav@tamu.edu, mannan@tamu.edu.

[2] Harold Vance Department of Petroleum Engineering, Texas A&M University, College Station, Texas 77843-3122, US, email: rashid.hasan@pe.tamu.edu.

[3]Hess Corporation, 1501 McKinney Street, Houston, TX 77010, US, email: skabir@hess.com.

With an increasing demand for hydrocarbons and other sources of energy, the methods for delivering such energy needs have become complex. In the case of the oil and gas industry, exploration and production have gone to deeper waters and deeper horizons than ever before, resulting in drilling several miles into the earth's core. Deeper well penetration means encountering increasingly high formation pressures and temperatures. For example, in deepwater drilling, well construction and production management deal with significant heat transfer between the produced fluids and its surroundings. The high-rate production and the attendant heat transfer demand close attention in terms of operational safety and potential environmental impact. One phenomenon that occurs in producing wells is sustained casing pressure (SCP), which presents a risk to both process safety and the environment.

In oilfield terminology, the casing is a steel pipe cemented to the surrounding formation. Multiple concentric casings are involved in a typical well construction. The casing immediately surrounding the production tubing is known as the production casing, which is most vulnerable to thermal and mechanical stresses and fluid leakage. The cement sheaths provide zonal isolation, support the casing, and protect it from corrosive fluids. Ideally, pressures monitored at the wellhead for multiple casings should change minimally after being bled off through a needle valve at the surface. When the pressure rebuilds after being bled off, that casing is said to exhibit SCP [1].

SCP occurs because of stresses that the system experiences owing to fluids production up the tubing. Besides the normal source of SCP, a poor cement job may allow hydrocarbon migration from the producing or other horizons and accumulation into one or more of the casings. The risk of not attaining well integrity can range from the activation of rupture discs to the blowout of hydrocarbons. Because of the safety and environmental dangers of SCP, regulations require remedial treatments for a casing string that exhibits significant SCP problems. Unfortunately, available methodologies to test for SCP are often qualitative and empirical in nature [2, 3], rendering them inapplicable in many situations. No practical systematic testing procedure has emerged to quantify SCP.

In this paper, we present a model for SCP that is rooted in the transport processes of the system. The model assumes gas leaking through the cement barrier, and then migrating up the annular-liquid column. It accounts for the fluid-temperature profile along the wellbore and liquid compressibility. Some robust assumptions about the transport processes allow us to write the governing equation as a first-order, linear-differential equation that is solved with appropriate boundary conditions. The resulting algebraic expression for the casing pressure rise as a function of time paves the way for all applications. A comparative study of the model's performance with field data suggests excellent agreement for wells where the casing pressure increase is owes gas influx into the annulus. The proposed model could form the basis for standardization of SCP testing. This study also offers some test procedures to assess SCP-related safety concerns.

References

1. Bourgoyne Jr., A.T.J., S.L. Scott, and W. Manowski, ***A Review of Sustained Casing Pressure(SCP) Ocurring on the OCS***, 2000, Final Report Submiteted to MMS.

2. Xu, R. and A.K. Wojtanowicz, ***Diagnosis of Sustained Casing Pressure from Bleed-off/Buildup Testing Patterns.*** SPE, 2001: p. Productions and Operations Symposium.

3. Wojtanowicz, A.K., S. Nishikawa, and R. Xu, ***Diagnosis and Remediation of Sustained Casing Pressure in Wells***, July 2001, MMS.

(45f) Evaluation of Particle Growth in an Electrochemical Fluidised Bed Reactor

Emmanuel Ehirim, Rivers State University of Science and Technology, Port Harcourt, Nigeria

The process of electrodeposition within electrochemical reactor results in the increase of particle sizes, which increases the volume of the bed. This work entails a mathematical model to describe the effect of this phenomenon on the behaviour of a three-dimensional fluidised bed electrode operating under the condition of electron transfer. The model incorporates the pseudo-stationary behaviour in the reactor since the bed increases in volume due to the particle size growth. The Orthogonal Collocation method was used to solve the developed differential equations. The overpotential distribution and the reaction rate within the reactor have been calculated. The principal results are the effects of the electron transfer operation on the superficial displacement with time. The effects of total applied current, electrode geometry and the bed porosity on the behaviour of the system have been analysed. The result shows a very gradual increase in the bed height.

(45g) Debottlenecking Gas Dryers and Desiccant Adsorbers

Benjamin A. Schmitt, Hydrocarbon Processing Industry (HPI) - Business Unit, Johnson Screens, Inc., New Brighton, MN and Michael Ekholm, Engineering, Johnson Screens, Inc., New Brighton, MN

Debottlenecking Gas Dryers and Desiccant Adsorbers

Though Maximization of the Useable Bed Volume with Johnson Screens' SSG^{TM}

Benjamin Schmitt, PE

Senior Mechanical Engineer, Technology

Michael Ekholm, PE

Applications and Product Design Manager

Johnson Screens Inc. St. Paul, MN

AIChE Spring Meeting, April 2013

13th Topical Conference

San Antonio, TX

Introduction:

The following presentation of the Shaped Support Grid (SSG^{TM}) releases additional supporting data and updates to the initial presentation given by Robert G. Norell at the 2010 AIChE Spring Conference.

Limitations on ethylene dryers and desiccant adsorbers can create bottlenecks in the production process which can lead to a need for increased dryer and adsorber capacity. Capital intensive investments may be necessary in order to eliminate the bottleneck which may include the addition of an additional vessel. With limits on the plot plan, this may not be a workable solution. Adding to the problem is the amount of time consumed in engineering, planning and construction, which can have negative effects on the operation of the ethylene unit. Increased desiccant volume and operational efficiency can increase the unit's production with minimal capital expenditures. Johnson Screens', patent pending, SSG^{TM} can add the increased process efficiency and cycle life necessary to extend the operating capability of the unit.

Extending the operation, though debottlenecking, can be achieved by maximizing the time before breakthrough of the media bed. There are three primary functions that maximize the breakthrough time:

Proper gas distribution in both normal operation and regeneration conditions which maximizes the bed utilization.

Increased molecular sieve or desiccant bed volume are achieved through the elimination of traditional bed supports comprised of support grids and beams. The support interface for the screen configuration also allows for thermal expansion typically present in vessels with a thermal cycle as part of the process. The design allows for thermal cycling without the risks of seal or structural failure inherent in other solutions.

Minimizing the bed volume consumed by the distribution and support system increases the net desiccant volume lost when utilizing alternative bed supports.

Johnson Screens ***SSG*** is designed to provide the industry with an alternative, cost effective, approach to increasing capacity. Supporting data and preliminary case studies will be depicted to quantify the benefits that are achievable through the implementation of the ***SSG***.

Conclusions:

The use of a Johnson Screens ***SSG*** provides a reliable immediate solution to reduce capacity bottlenecks. The ***SSG*** increases the bed volume, when compared to traditional bed support arrangements comprised of grids and beams. The increase in desiccant volume will allow for increases in feed flow rate and/or adsorbtion cycle time. The ***SSG*** maximizes the flow distribution during desiccant regeneration, maximizing the utilization of the complete cross section of the vessel. This solution can be implemented in both new and existing vessels, providing consistency in design between planned increases in performance of current vessels and future expansions.

The modular design of the ***SSG*** allows for easy installation and maintenance. Connection of the ***SSG*** to the vessel is performed within the outlet nozzle, eliminating all welding on the pressure vessel and eliminating any need for recertification of the pressure vessel after installation. The design allows for ease of dismantling for inspection of the vessel head during routine maintenance and inspections. The low profile will also reduce interference with any required media exchanges. The design package of the ***SSG*** provides the end user with a holistic solution that fulfills debottlenecking needs with limited capital expenditures.

Biography of Speaker:

Benjamin Schmitt is a Senior Mechanical Engineer in Johnson Screens' Hydrocarbon Processing Industry Technology Group. Ben graduated from North Dakota State University in 2004 with a Bachelor of Science degree in Mechanical Engineering. He currently working toward his Master's in Business Administration from Hamline University. Ben is member of AIChE and is a registered professional engineer in the state of Minnesota.

AMACS
PROCESS TOWER INTERNALS

AIChE 2013 Spring Meeting Poster board
Design Considerations for effective
liquid removal In Suction Drums

Presented By:
Kanti Patel, PE

Compressor down again?

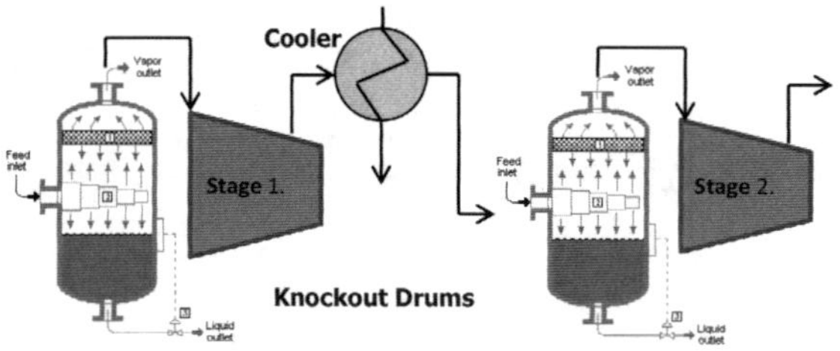

Is liquid carryover the culprit?

<u>We have heard it all before</u>:

-Premature compressor shutdown.

-Production time lost.

-High yearly compressor maintenance costs.

-Impeller damage due to pitting corrosion.

-Corrosion due to salt deposits.

AMACS
PROCESS TOWER INTERNALS

Common Design Problems with Suction Drums which promotes liquid carryover

1. Choosing the correct mist eliminator (M.E. equipment)
2. Knowing where to install the M.E. in a suction drum.
3. Consider a high mist loading in the gas when selecting your M.E.
4. Fouling considerations.
5. M.E. disengaging space, bad gas distribution & liquid slug in the feed.
6. One micron and great droplets, versus submicron mist removal.
7. High viscosity liquids.
8. Dislocated M.E. due to pressure surges.
9. Through put exceeds original design capacity.
10. Capacity range too large for suction drum.

AMACS
PROCESS TOWER INTERNALS

Choosing the correct Mist eliminator equipment

- Proper sizing

+ Use Souders-Brown equation to find optimum velocity. $$\mathbf{Vd = K} \cdot \sqrt{\frac{Pl}{Pg} - 1}$$

+ Select the correct vapor velocity factor K. based on the operating pressure, and type of equipment.
    - Lower K-factor 0.01 for every 100 psi
    - K factor to select when using:

Mesh Pad	K= 0.1 - 0.55
Chevron Vane	K= 0.45 - 0.65
MultiPocket® Vane	K= 0.75 – 1.0

AMACS
PROCESS TOWER INTERNALS

Choosing the correct Mist eliminator equipment cont.

- Selection based on droplet size.

Table 2. Droplet sizes (water in air) typically captured with 99.9% efficiency by mist eliminator elements of various types

Element type	Size range microns
Fiber candles or panels	0.1 and larger
Mesh with co-knit yarn	2.0 μm and larger
0.006-inch knitted mesh	5.0 μm and larger
0.011-inch knitted mesh	10 μm and larger
AMACS MultiPocket ® Vane	10 μm and larger
Conventional vane arrays	15 μm and larger

AMACS
PROCESS TOWER INTERNALS

Where to install the M.E. in the suction drum

- Improperly installed M.E. can cause severe liquid carryover problems.
- General accepted spacing guidelines.

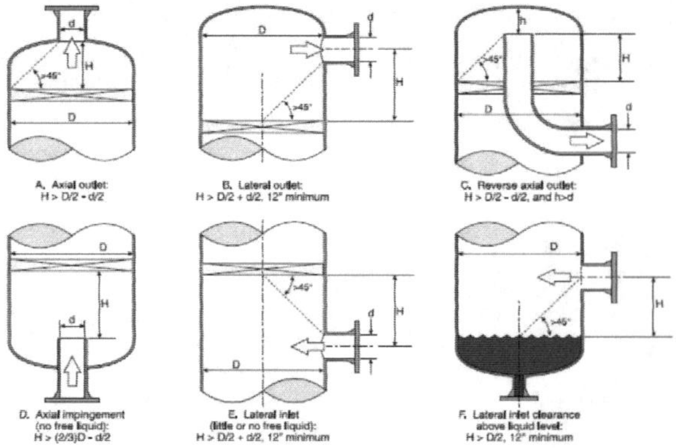

AMACS
PROCESS TOWER INTERNALS

Consider high mist loading when selecting your M.E. equipment

+ High mist loading in the gas stream can flood your M.E. equipment and cause liquid carryover.

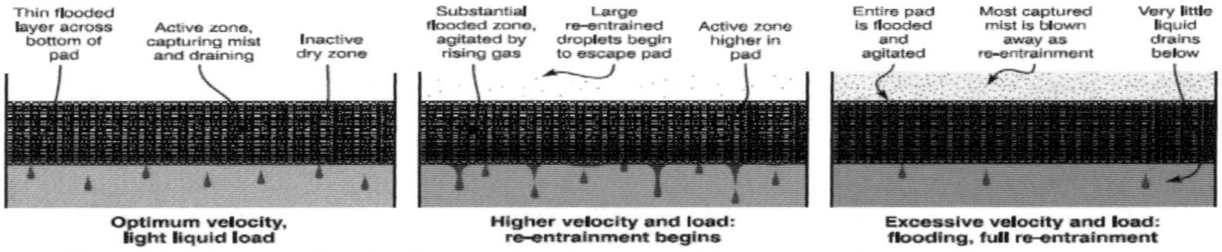

Figure 4. Envisioning mesh pad performance degradation as liquid and gas loads increase (vertical cross-sections)

AMACS PROCESS TOWER INTERNALS

Consider high mist loading when selecting your M.E. equipment cont.

+ Limited drainage capabilities
  - mesh M.E. <0.5 gal/FT2/min
  - Vane type M.E. <2 gal/FT2/min

+ Select a combination of vane & mesh mist eliminators

AMACS PROCESS TOWER INTERNALS

Fouling Applications

+ Fouling of mesh mist eliminator will cause liquid carryover.

+ For mild fouling install a water spray system below the mist eliminator for occasional inline cleaning (based on Delta-P)

+ For consistent fouling problems, install a Vane mist eliminator upstream of the mesh mist eliminator.
    - Vanes remove solids and bulk liquids therefore, protecting the downstream mesh mist eliminator

AMACS
PROCESS TOWER INTERNALS

Fouling Applications cont.

+ For consistent fouling problems where the solids are "sticky" install a water spray system below the vane mist eliminator.

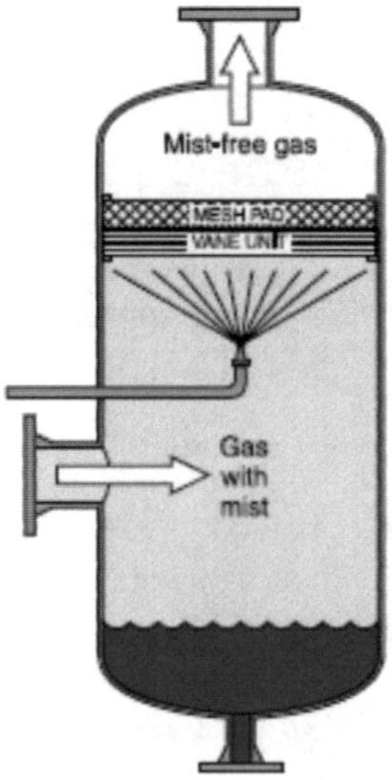

AMACS
PROCESS TOWER INTERNALS

Bad Gas distribution, limited disengaging space and liquid slugs.

+ Bad gas distribution, limited disengaging space in a suction drum can cause severe liquid carryover.

+ CFD analysis can assist in locating high velocity areas.

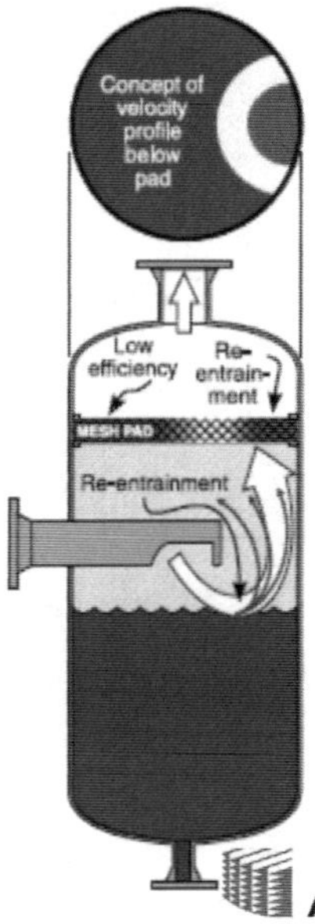

AMACS
PROCESS TOWER INTERNALS

Bad Gas distribution, limited disengaging space and liquid slugs cont.

+ Use inlet diffuser to force uniform gas distribution i suction drum.

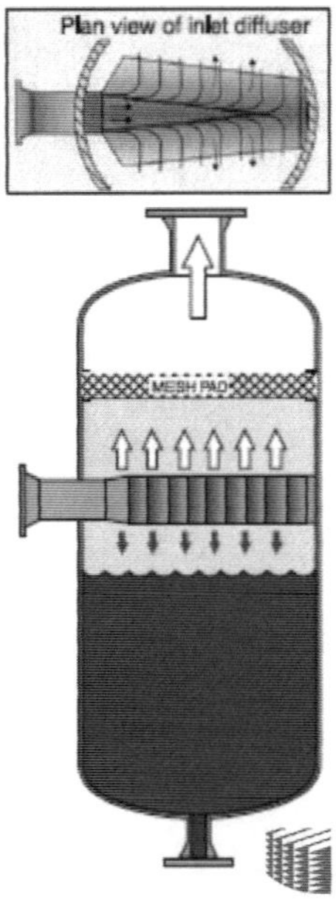

AMACS
PROCESS TOWER INTERNALS

Inlet Feed Device

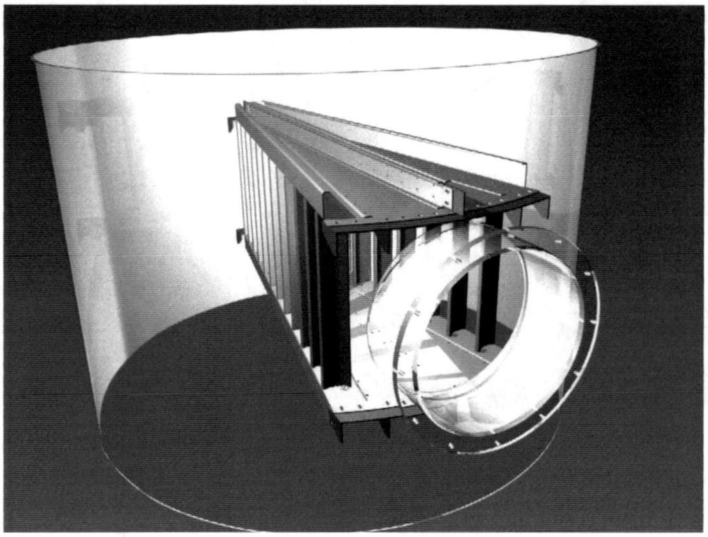

AMACS
PROCESS TOWER INTERNALS

Inlet Feed Device

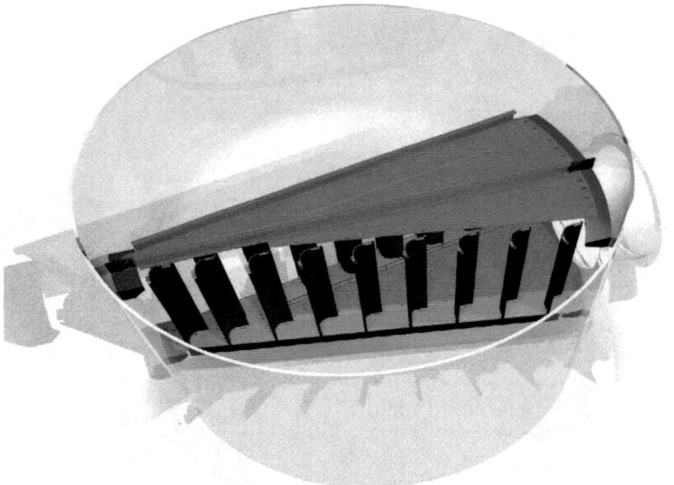

AMACS
PROCESS TOWER INTERNALS

Bad Gas distribution, limited disengaging space and liquid slugs cont.

+ Use a perforated distribution plate downstream of the M.E. for additional distribution assistance.

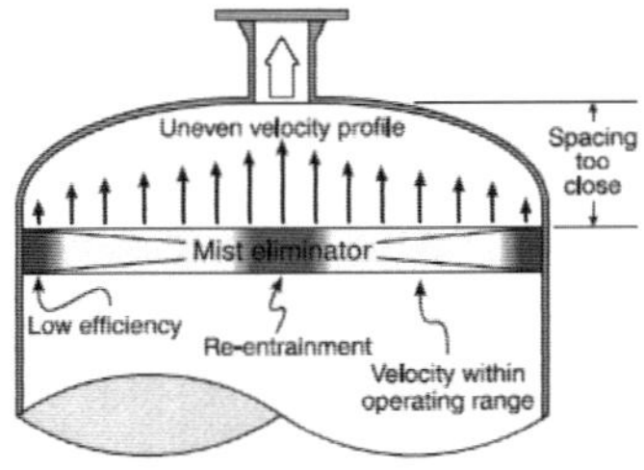

Figure 7. Example of mist eliminator performance degradation due to uneven velocity profile

+ Inlet diffusers also handle incoming liquid slug which otherwise can cause flooding of the Mist eliminator.

AMACS
PROCESS TOWER INTERNALS

Know what mist size droplets You are dealing with

+ Know if your mist is > 1 micron or submicron.
+ When gas travels from A to B, condensation in the pipeline will cause a fine mist (fog or submicron droplets)

  Examples:

  - Incoming gas from an offsite location
  - Fuel gas to furnaces/heaters with low NOx burners

AMACS
PROCESS TOWER INTERNALS

Know what mist size droplets You are dealing with cont.

+ Submicron mist is not captured by conventional M.E.

+ Submicron mist requires a coalescing filter, typically
installed in a two stage filter separator.

AMACS
PROCESS TOWER INTERNALS

Know what mist size droplets You are dealing with cont.

<u>1</u>

1ᶜ stage; submicron droplets are coalesced to > 10 micron droplets.

2ᶜ stage; greater than > 10 micron droplets are ejected from the coalescing filter and captured by a vane, which captures droplets > 8 micron.

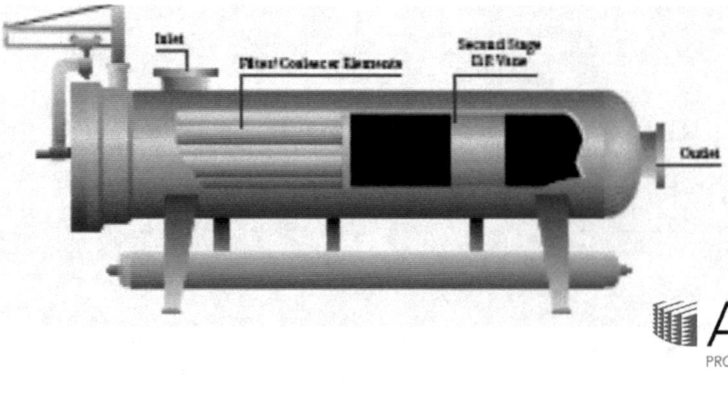

AMACS
PROCESS TOWER INTERNALS

High viscosity liquid as mist

+ Viscous liquids drain slower out of the M.E. and could cause localized flooding.
+ Use a mesh mist eliminator with an open drainage layer on the bottom to improve liquid drainage.
+ Use a lower K- value (lower face viscosity) for the M.E. To allow additional drainage time.
+ Use a vane type M.E. upstream of the mesh mist eliminator, to remove the bulk of the viscous liquid, so the M.E. can effectively handle the residual liquids.

Dislocated M.E. due to System pressure surges

+ M.E. sections can be dislodged into the downstream equipment, which then creates massive carryover.
+ Standard M.E. equipment often gets installed and secured to the support ring with tie wire.

Dislocated M.E. due to System pressure surges cont.

+ Where system pressure surges are common, there are more secure ways to install the M.E.

  - dual support rings
  - through bolts
  - install a pressure relief door

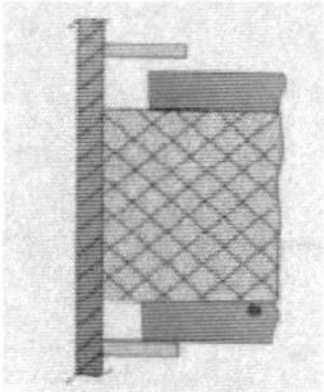

Dual support rings

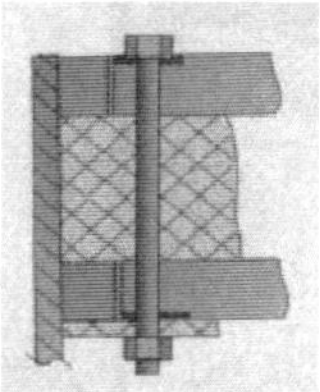

Through bolts

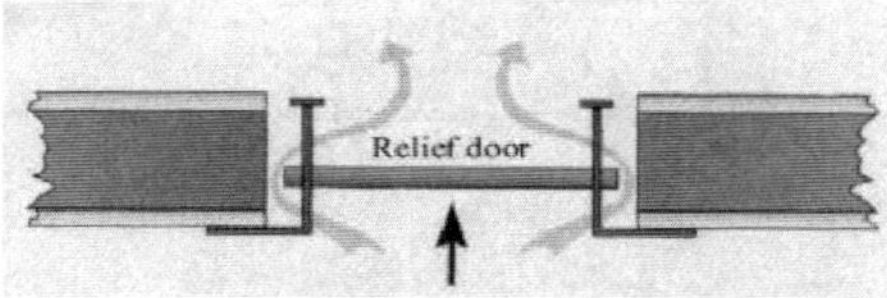

AMACS
PROCESS TOWER INTERNALS

When throughput exceeds original capacity

+ There are various solutions to increase capacity in an existing vessel.

  - Patented High capacit
    mist eliminator 0% - 1

AMACS
PROCESS TOWER INTERNALS

When throughput exceeds original capacity cont.

-Patented MISTFIX™
cylindrical mist eliminator 0% - 25%

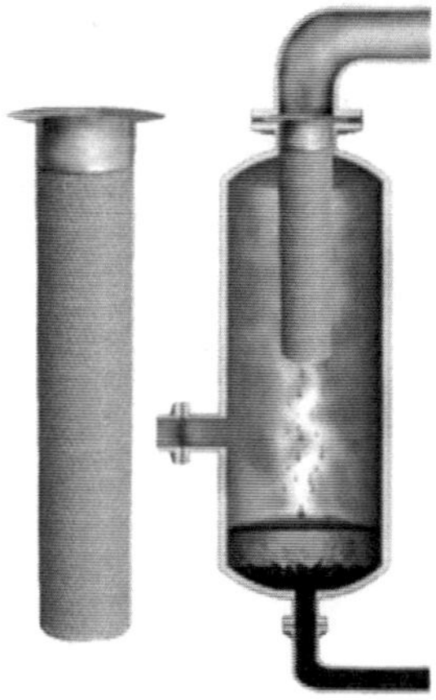

When throughput exceeds original capacity cont.

- High capacity Patented
MultiPocket® vane.
0%-50%

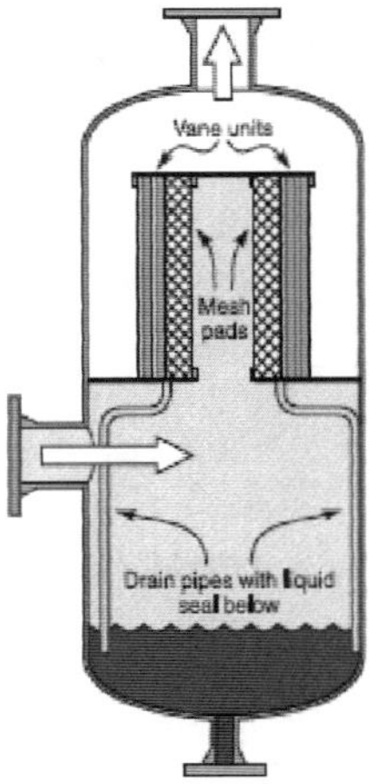

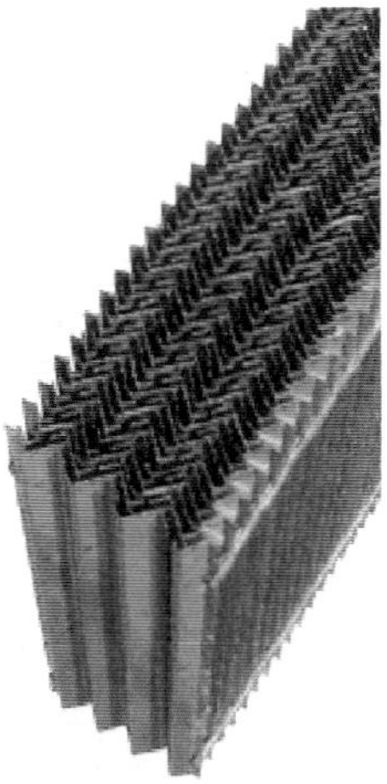

AMACS
PROCESS TOWER INTERNALS

When throughput exceeds original capacity cont.

+ MistFix™ is installed through the gas outlet, all other M.E. equipment comes in sections, installed through the manway.
+ With increased capacity comes increased bad gas distribution.
  - Use an inlet diffuser.
+ To avoid welding new support rings and housing for the high capacity new equipment, use double expansion rings, a compression fit.

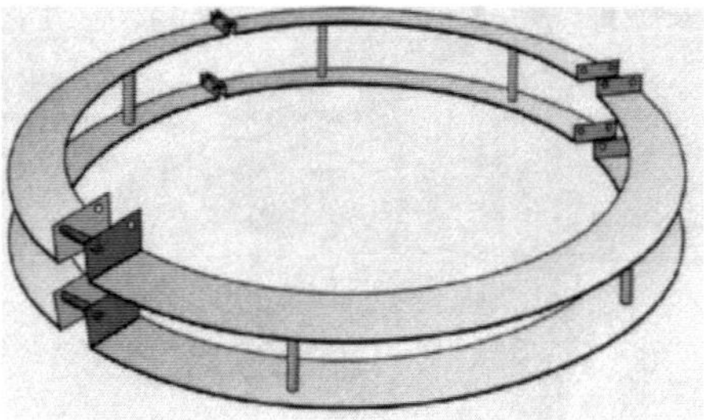

AMACS
PROCESS TOWER INTERNALS

Support Rings

AMACS
PROCESS TOWER INTERNALS

Support Rings

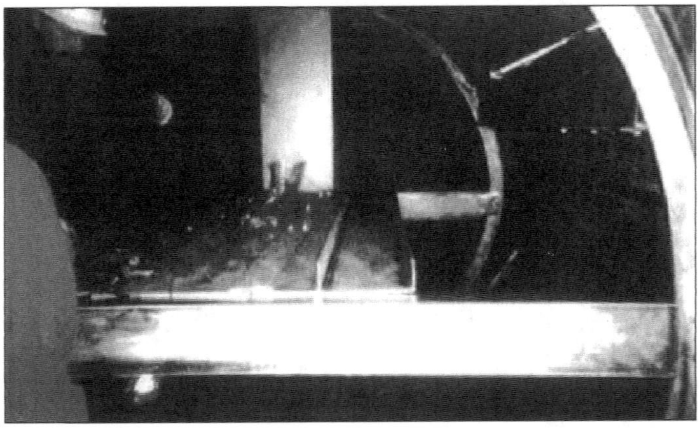

AMACS
PROCESS TOWER INTERNALS

When the capacity range is too large for a suction drum

+ Typical M.E. equipment will operate between
  30% – 110% capacity.

+ When a suction drum often operates below 30% liquid carryover may occur.

AMACS
PROCESS TOWER INTERNALS

When the capacity range is too large for a suction drum Cont.

To facilitate a capacity range of 10% - 110% consider a 2 phase M.E. in the same drum.

- one that operates 30% - 1
- second that operates between 10% - 30%

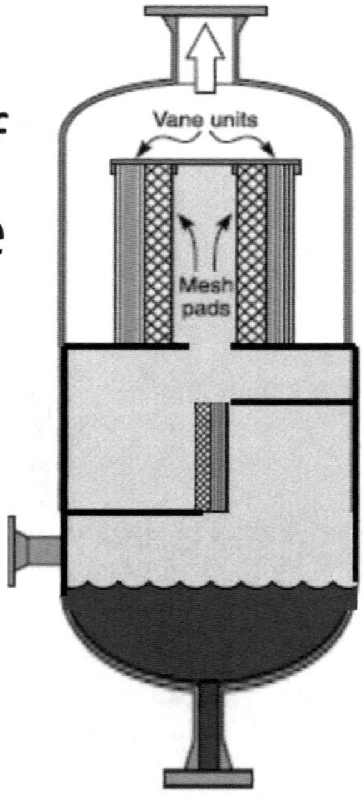

(45i) Renewable Feedstock for Steam Crackers: Catalytic Upgrading of Crude Tall Oil (CTO) Into Bio-Naphtha

Jinto Anthonykutty, Juha Linnekoski, Antero Laitinen and Ali Harlin, VTT Technical Research Centre of Finland, Espoo, Finland

Renewable feedstocks from bio resources are better alternatives to petroleum derived naphtha for the production of olefins and aromatics by steam cracking or reforming.[1, 2] Mostly, these renewable feedstocks (bio-naphtha) rich in n-alkanes are produced by catalytic upgrading methods via hydrotreating/hydroprocessing; in which oxygenates are removed from bio-oil by various deoxygenation processes. At present, hydrotreated vegetable (HVO) oil is produced in many countries on industrial scale using existing petroleum refinery methods[3]. However, employing higher priced vegetable oils could be a major concern in these processes and the main challenge is therefore finding an inexpensive and non-food chain affecting feedstock. Tall oil, obtained as a by-product from Kraft pulping process, is abundant and cheap;and hence, meets the criteria of a low cost starting material for upgrading[4]. Moreover, the utilization of tall oil in conventional oil refineries could serve as an eco-friendly method to convert bio-wastes into bio-naphtha[1].

In the present research approach, catalytic upgrading (hydrotreating) of crude tall oil (CTO) into bio-naphtha suitable for a steam cracker is discussed. CTO, which contains mainly fatty acids (42.9 wt. %) and resin acids (25.7 wt. %) was obtained from Stora-Enso Kaukopää (Finland) mill and used as received. A commercial NiMo catalyst was employed and the experiments were conducted in a continuous, down flow, fixed bed reactor at different process conditions (WHSV 1-3h^{-1}, Pressure 5 MPa and Temperature 350 - 450 ^{o}C). After the stabilization period (6 h.), the liquid sample was collected and separated into organic and aqueous phases.The product distribution in organic phase was analyzed with a Shimadzu GC/MS equipped with a Zebron MS 30m × 0.25 mm × 0.25μm column.

This current approach reports that CTO produces mainly paraffins, cycloalkanes and aromatics during upgrading process. Paraffinic hydrocarbons mainly comprise n-octadecane (20-25 wt. %) and n-heptadecane (15-20 wt. %), are obtained in maximum yield (55- 60 wt. %) in low temperature reactions. The yield of paraffinic hydrocarbons increased as a function of increasing space time in low temperature reactions. 18-Norabietane and other mono-, di- and tri- cycloalkanes are obtained from resin acids in CTO, and these cyclic structures are converted to aromatics such as naphthalene and phenanthrene in high temperature reactions (>400 ^{o}C). Stigmastane, apparently produced from sterols in CTO are also present as a major three ring component in the product stream. Additionally, gaseous products obtained and the yield increased as a function of temperature. The trend observed with the distribution of gaseous products reveal that decarboxylation and cracking reactions are favored at higher temperatures, while hydrogenation and hydrodeoxygenation (HDO) are suppressed.

The elemental composition of CTO shows that it contains relatively high sulfur content (1800 mg/kg), could be useful to keep the catalyst's activity with time on stream (TOS). However, the considerable amount (87 mg/kg) of alkali metals (Na, K, Ca and P) in CTO might be critical for the catalyst activity with TOS, will be discussed in this contribution.

The product composition obtained with CTO is assessed in comparison with compositions and characteristics of refinery naphtha feeds, and will be discussed in detail. Furthermore, this contribution will discuss the achieved deoxygenation rate and success criteria for the product composition of upgraded CTO.

References

1. Gornay, J.; Coniglio, L.; Billaud, F.; Wild, G. Steam Cracking and Steam Reforming of Waste Cooking Oil in a Tubular Stainless Steel Reactor with Wall Effects. ***energy&fuels***, **2009**, 23(11), 5663-5676.
2. Pyl, S.P.; Schietekat, C.M.; Reyniers, M.F.; Abhari, R.; Marin, G.B.; Van Geem, K.M. Biomass to olefins: Cracking of renewable naphtha. ***Chemical Engineering Journal***, **2011**, 176-177, 178-187.
3. Sunde, K.; Brekke, A.; Solberg, B. Environmental Impacts and Costs of Hydrotreated Vegetable Oils, Transesterified Lipids and Woody BTL- A Review. ***Energies***, **2011**, 4(6), 845-877.
4. Coll, R.; Udas, S.; Jacoby, W.A. Conversion of the Rosin Acid Fraction of Crude Tall Oil into Fuels and Chemicals. ***energy&fuels***, **2001**, 15, 1166-1172.

AIChE Paper Number 45j
High Efficiency Ethylene Cracking Furnaces

Rajaram Ramesh

Product Development Manager - Ethylene
Technip
Netherlands

Marco W. M. van Goethem, PhD

Product Development Manager – Modeling and Testing
Technip
Netherlands

25th Ethylene Producers Conference
AIChE 2013 Spring National Meeting
San Antonio, Texas, April 29 – May 3, 2013

Abstract

Environmental and economic pressures generate a driving force to produce products more efficiently. A direct effect of more efficient processes is that the emission of green house gases, particularly carbon dioxide decreases. In households nowadays "high efficiency furnaces" are available with efficiencies in the range of 94-97%. This paper describes the outcome of a feasibility study on applying the high efficiency furnace concept to steam cracking furnaces. The high efficiency is achieved by enhanced heat recovery from the flue gases before they are released to the atmosphere at ca. 65°C.

Technip studied the application of this concept for cracking furnaces based on gas as well as liquid feedstocks. Natural gas was used as basis for the energy price. Since energy prices have a strong impact on project economics, the evaluations were done for an EU average natural gas price and a US natural gas price.

Two alternative heat sinks have been considered: preheating of combustion air and preheating of cold hydrocarbon feed. Both are viable for liquid cracking furnaces. In the case of gas cracking furnaces feed preheating results in a pinch and this makes it not viable.

Capital costs are based on Technip's cost database and quotations from air preheater vendors. Special attention has been paid to the material selection of the affected convection section banks.

This study shows that high efficiency ethylene cracking furnaces are economically viable. For Europe the study indicates an attractive payback period. In the US the payback period is longer due to the lower energy prices. If energy prices increase in the future, the concept's economic viability will become better. Inclusion of CO2 credits will make the payback period more attractive.

Furnace Efficiency

In this paper furnace efficiency has been calculated as net thermal efficiency as interpreted by API560[1)]. Furnace efficiency is determined by drawing an envelope around the furnace and including in it the radiant section, convection section and any heat exchanger equipment that recover heat from the flue gas (air preheater, economizer, steam superheater, etc.). The efficiency is the ratio of the useful heat absorbed to the fired heat of the fuel and any other external heat inputs. Ambient temperature is a reference temperature for efficiency calculations and any heat inflow or outflow above this reference is taken into consideration as a heat input or heat loss.
Furnace efficiency can be determined as follows:

$$\frac{heat\ absorption\ in\ radiant\ section\ +\ convection\ section}{(firing\ rate\ \times\ fuel\ heating\ value) + sensible\ heat\ of\ fuel\ and\ air\ above\ ambient\ temp. + other\ external\ heat\ inputs}$$

Alternately, $\text{Furnace efficiency} = 100\% - \text{Total heat loss}$, where:

$$\text{Total heat loss} = \frac{stack\ loss\ +\ radiation\ heat\ loss}{(firing\ rate\ \times\ fuel\ heating\ value) + sensible\ heat\ of\ fuel\ and\ air\ above\ ambient\ temp. + any\ external\ heat\ inputs}$$

In a furnace where fuel and combustion air enter the furnace at ambient temperature and there are no other external heat inputs, the first expression can be simplified to the following:

$$Furnace\ efficiency = \frac{heat\ absorption\ in\ radiant\ section\ +\ convection\ section}{fuel\ firing\ rate\ \times\ fuel\ heating\ value}$$

Please note that net thermal efficiency, as defined by API 560, is based on the lower heating value of the fuel.

Study objectives

Technip undertook this study with the goal to reduce fuel firing in cracking furnaces by at least two percent. Although this figure does not look very ambitious, it must be remembered that since efficiencies in current designs are already in the mid nineties (upper part of the technology S-curve), further improvements will not be easy and can only be incremental.

Since stack loss is the main source of inefficiency (radiation heat loss is comparatively smaller and more difficult / expensive to reduce) it was decided to make stack loss reduction as the main focus of the study.

By a relatively straight forward calculation it can be shown that the two percent efficiency rise target would require the flue gas to be cooled to approximately sixty or seventy degrees Celsius.

The main challenge in reducing stack temperature to this level is finding an appropriate heat sink for the recovered heat. In a cracking furnace there are two suitable heat sinks, combustion air and hydrocarbon feed, and both were studied. The following sections discuss the results for liquid furnaces and gas furnaces.

Basis of Study

We have done the study on a naphtha furnace and an ethane furnace. We have chosen these two feedstocks since a large number of furnaces operate on these two feeds and we can draw conclusions that can be applied approximately to other liquid and gas feeds. The study has been performed on the following bases:

- The reference liquid and gas furnace designs used in this study are based on actual furnaces from Technip's past projects. The furnaces were simulated using Technip's proprietary tool SPYRO which is used in the design and simulation of cracking furnaces.
- The budgetary capital cost need to achieve the target furnace efficiency was calculated as an incremental cost. It was estimated from vendor quotes and Technip's internal cost database.
- The cost estimates are for a new project, not a revamp, and are for a West European location.
- Operating costs are based on cost of base utilities and derived utilities. Base utilities are natural gas and demineralized water. These are taken from market data.
- Since natural gas prices are very different in different parts of the world, the operating costs and economic analysis have been done for European and United States gas prices.
- EU natural gas price was fixed at €25.1 /MWh of higher heating value (HHV) which is equivalent to €28.0 /MWh of lower heating value (LHV). The prices are based on EUROSTAT data for 1st half of 2010 and were obtained from the publications of the Department of Energy & Climate Change in the UK. They correspond to the EU-15 average for Large Industrial Consumers (i.e. 2.13 pence/kWh HHV).
- US Natural gas prices were obtained from the US Department of Energy. The average value was 6 \$/MM BTU (HHV) or 17 €/MWh (LHV) for 1st half of 2010. The average HHV used is 1029 BTU per cubic foot or 1.029 MM BTU per thousand cubic feet.
- Cost of demineralized water was assumed at € 2.00 /ton.
- The prices of other utilities (boiler feed water, steam and electricity) have been calculated from those of the two base utilities. Electricity was priced assuming 35% overall efficiency of power generation.
- The economic calculations do not include any CO2 credit for the fuel saved.
- Assumed exchange rate: 1 US dollar = 0.75 euros
- The plant is assumed to operate 8400 hours per year

The resultant utility prices used in the study are summarized below.

UTILITY COSTS

		EU basis	US basis	
FUEL as NATURAL GAS	Base	**28.0**	**17.0**	€ / MWh
DEMINERALISED WATER	Base	2.0	1.5	€ / ton
VHP STEAM (net value)	Derived	23.7	14.4	€ / ton
MP STEAM (net value)	Derived	13.9	8.5	€ / ton
LP STEAM	Derived	9.7	5.9	€ / ton
ELECTRICAL POWER	Derived	80.1	48.6	€ / MWh

Study Results

Naphtha Furnace using Air Preheat

Combustion air can be preheated in a plate type air preheater (APH). These are cross flow type exchangers and have low pressure drops. Combustion air is at a lower temperature than any of the process or utilities streams entering the convection section; therefore, from the standpoint of efficiency, APHs are ideally suited to be the last heat recovery section before the stack.

Fig 1 shows the Base Case Naphtha furnace while Fig 3 shows the same furnace with an APH. Please also refer to Table 1 for a performance comparison of the APH design with the Base Case naphtha furnace. The following points are of interest:

1) The overall efficiency of the furnace will improve from 94.4 to 97.0% for the APH option. This is mainly due to the lower stack temperature but also partly due to the smaller excess air, which is explained below.

2) This option will require a forced draft combustion air fan to overcome the pressure drop in the APH. As a result the furnace operation will change from induced draft mode to balanced draft. The availability of air in forced draft will enable more accurate control of the excess air and it will be possible to operate the furnace with lower excess air (5% instead of 8%). This leads to a lower flue gas flow, a lower stack loss and therefore better efficiency.

3) There will be a drop of 4.8% in fuel firing which is greater than the efficiency increase. This is because of the lower absorbed heat. The furnace is designed and controlled in such a way as to supply all the required process heat and to absorb the surplus heat into utilities to the extent possible. When the furnace operates more efficiently and the heat demanded by the process media remains unchanged, as in the APH case, there will be less surplus heat available for VHP steam. As a result the VHP steam generation will go down from 79.7 to 76.0 t/h.

4) The heat transfer surface in some parts of the APH will be below the water dew point of the flue gas (around 60°C). This can lead to water condensation and create two possible issues: 1) accumulation of water in the APH and 2) corrosion due to the presence of liquid water, oxygen and carbon dioxide.

   Two measures have been included in the design to counter these problems:
   - Placement of unit such that flue gas flows downward to aid easy draining
   - Use of corten steel to resist corrosion

5) We have evaluated the impact of the air preheat option on NOx emissions using API 535[2)]. The rise in air temperature and the drop in excess air will have opposing effects on the NOx production. The net effect is that the NOx concentration will be close to the Base Case.

   If a DeNOx unit is present, as is often the case in the US, the APH option may help to slightly extend the life of the DeNOx catalyst. This is because the total NOx load will be smaller than the Base Case due to the smaller flue gas flow.

Table 2 summarizes the results of the economic analysis of this option. We have used payback period to compare different options. Payback period is the ratio of the incremental capital cost above the Base Case CAPEX divided by the savings in yearly operating cost.

1) The capital cost (CAPEX) of the APH option includes the cost of the APH, the combustion air fan, extra ducting and structural steel. The operating cost (OPEX) savings is the savings in furnace fuel but this must be offset against the value of the lost VHP steam production (which notionally must be made up in a utility boiler elsewhere) and the additional electricity for the two fans.

2) It can be seen that the payback period is three years based on European gas prices and five years based on American gas prices.

Naphtha Furnace using Feed preheat

Naphtha is normally stored at ambient temperature and is preheated to 60°C using quench water before it is routed to the furnaces. Most cracking furnaces are equipped with a Feed Preheater (FPH) bank in the convection section. If flue gas is utilized to preheat naphtha from ambient conditions the stack temperature can be lowered to the desired range of sixty to seventy degrees Celsius. Till now designers and operators have not taken full advantage of this mainly because of the higher CAPEX and fear of dew point corrosion.

One of the consequences of this feature will be that the quench water duty previously absorbed by the naphtha will now shift to the air coolers in the quench water circuit. The additional electrical power that the air cooler fans will require will be about 10 KW per furnace.

Please refer to Table 3 for an overview of the performance of a naphtha furnace with and without an extended feed preheater. Please note that the Base Case naphtha furnace is the same as the one considered in the APH option.

1) The overall efficiency of the furnace can be improved from 94.4 to 96.0% for this option.

2) Unlike the APH option a combustion air fan will **not** be required. The furnace will continue to operate in induced draft mode.

3) The fuel firing and the VHP steam generation will drop marginally.

4) The upper tube rows of the FPH will be below the water dew point of the flue gas and some condensation will occur. As a precaution we have considered that the top four rows will be of stainless steel.

5) The flue gas NOx concentration will be unaffected by this option. However the total hourly NOx emissions and load on the DeNOx unit, if any, will decrease to the same extent as the fuel firing, which is around 0.6%.

Table 4 summarizes the results of the economic analysis of this option.

1) The capital cost (CAPEX) of this option covers the incremental cost of the extended FPH, extra ducting and structural steel. The OPEX savings is the reduction in furnace fuel after adjusting for the reduction in VHP steam production and the additional electricity for the ID fan and the Quench Water Air Cooler fans.

2) In the Base Case the naphtha feed piping between the external feed preheater and the convection section is insulated for heat conservation. In the FPH option naphtha will be fed at ambient temperature to the furnace and so these lines do not have to be insulated. This will yield some capital cost saving but this has not been taken into account in our analysis.

3) The payback period excluding CO2 credit will be three and a half years with European natural gas and five and a half years with US natural gas.

Ethane Furnace using Air Preheat

Fig 2 shows the Base Case Ethane furnace that was used in the study. The APH arrangement on an ethane furnace will be quite similar to the one on the naphtha furnace. Table 5 shows the impact that the APH would have on the furnace performance. The main points are as follows:

1) The overall efficiency of the furnace will improve from 91.7 to 96.5% for the APH option. This is mainly due to the lower stack temperature but also partly due to the smaller excess air.

2) In this furnace also the APH will require a forced draft combustion air fan. This will allow the excess air to be reduced from 8% to 5% which will result in lower stack losses and a better efficiency.

3) The fuel firing will go down by 11.0%, from 8.2 to 7.3 t/h. The absorbed heat will drop from 116.8 to 110.6 MW. As a result of these the VHP steam generation will go down from 97.0 to 88.8 t/h.

4) A portion of the APH will operate below the flue gas water dew point. Consequently the unit must be designed to prevent water accumulation and corrosion.

5) The impact on NOx was evaluated here also. The rise in air temperature will have a greater impact on NOx than the fall in excess air. We estimate that as a result the flue gas NOx concentration will increase slightly (around two percent). This can be a cause for concern if the furnace does not have a DeNOx unit and is operating close to the limits imposed by authorities. However since the total firing will go down, the hourly NOx production will drop. Plants can consider discussing with authorities to obtain a higher allowable NOx limit,

   If the furnace has a DeNOx unit, the chemical load on the unit will reduce, which should help to obtain a longer catalyst life.

The operating cost savings and payback for this option have been summarized in Table 6.

1) The capital cost includes the cost of the APH, the combustion air fan, extra ducting and structural steel. The operating cost savings is the savings in fuel firing with an adjustment for the value of the reduced VHP steam generation and the extra electricity for the ID fan and FD fan.

2) The payback period with natural gas priced at European rates will be one and a half years. The corresponding figure for US gas prices will be three years.

Ethane Furnace using Feed Preheat

Using feed at ambient temperature does not significantly reduce the stack temperature in a gas furnace. This is because the feed preheat curve has a steeper slope than the flue gas cooling curve, causing a pinch on the hot side of the Feed Preheater. Fig 4 shows the heating and cooling curves of ethane, boiler feed water and flue gas in the convection section. The hatched shape on the plot shows the maximum additional heat duty that can be recovered into feed, over and above the Base Case. The plot clearly shows that when the ethane feed temperature is lowered from the base value of 55°C to 30°C, the stack temperature will drop from 170°C to about 150°C. The flue gas cannot be cooled any more than this modest amount.

For this reason the feed preheat option was not investigated any further.

Conclusions and Summary

The key results from the different study options are listed together below.

Cracking furnace feed	Naphtha		Naphtha		Ethane	
Cold sink	Combustion air		Hydrocarbon feed		Combustion air	
	EU nat. gas	US nat. gas	EU nat. gas	US nat. gas	EU nat. gas	US nat. gas
OPEX savings	€ 0.60 mill	€ 0.36 mill	€ 0.17 mill	€ 0.10 mill	€ 1.27 mill	€ 0.77 mill
Payback time	3 years	5 years	3.5 years	5.5 years	1.5 years	3 years
Salient features of each option	Small number of additional equipment		No additional equipment		Small number of additional equipment	
	Higher CAPEX		Lower CAPEX		Higher CAPEX	
	More OPEX savings		Less OPEX savings		Best OPEX savings	
	Slight effect on NOx concentration		No effect on NOx		Around 2% higher NOx concentration	

All three scenarios have attractive payback periods and will be worth considering in high energy cost countries such as Europe and Asia. In US and other energy rich nations, the economic incentive will be much lower. In the US, where ethane cracking is common and increasing even more, improvement of furnace efficiency with air preheat can be worth considering as the payback time is less than 3 years.

If the options are compared against each other, in our opinion the Air preheating option has more potential, irrespective of feedstock.

The Feed Preheating option in liquid furnaces is cheaper but it also gives proportionally modest savings. Moreover, as earlier mentioned, the feed preheat duty is being shifted from one waste heat source (quench water) to another (flue gas). This option will not lead to a net improvement or cash savings in the plant as a whole unless there is a shortage of quench water and there are other, better uses for the saved quench water duty

Acknowledgements

The authors wish to express their thanks to their colleagues Mr Ashok Laghate, Ms Susan Cooper, Mr Sachin Singare and others in Technip Benelux who contributed to this study.

References

1) API 560: Fired Heaters for General Refinery Service, 4th edition

2) API 535: Burners for Fired Heaters in General Refinery Service, 2nd edition

Table 1
Naphtha Furnace with Combustion Air Preheating – Performance Summary

		Base Case	Air Preheat Case
Naphtha feed	t/h	60.8	60.8
Ambient conditions	-	15°C, 85% R.H.	15°C, 85% R.H.
Air temp. to burners	°C	15	77
Stack temperature	°C	108	61
Excess air	%	8	5
Absorbed heat	MW	113.2	110.4
Overall efficiency	%	94.4	97.0
Fuel flow	t/h	8.4	8.0
Fired heat	MW	119.9	113.9
VHP steam production	t/h	79.66	76.01
NOx concentration		Base	Close to Base

Table 2
Economics of Naphtha Furnace with Combustion Air Preheating

		Air Preheat Case (EU location)	Air Preheat Case (US location)
Savings in fuel firing	million €/yr	1.42	0.86
Cost of lost VHP steam capacity	million €/yr	0.73	0.44
Cost of additional electricity	million €/yr	0.10	0.06
Net OPEX saving	million €/yr	0.60	0.36
Payback	Years	3	5

Table 3
Naphtha Furnace with Feed Preheating – Performance Summary

		Base Case	Feed Preheat Case
Naphtha feed	t/h	60.8	60.8
Feed temperature	-	60°C	30°C
Combustion air temp.	°C	15	15
Stack temperature	°C	108	71
Excess air	%	8	8
Absorbed heat	MW	113.2	114.3
Overall efficiency	%	94.4	96.0
Fuel flow	t/h	8.4	8.35
Fired heat	MW	119.9	119.1
VHP steam production	t/h	79.66	79.63
NOx concentration		Base	Base

Table 4
Economics of Naphtha Furnace with Feed Preheating

		Feed Preheat Case (EU location)	Feed Preheat Case (US location)
Savings in fuel firing	million €/yr	0.19	0.12
Cost of lost VHP steam capacity	million €/yr	< 0.01	< 0.01
Cost of additional electricity	million €/yr	0.01	< 0.01
Net OPEX saving	million €/yr	0.17	0.10
Payback	Years	3.5	5.5

Table 5
Ethane Furnace with Air Preheating – Performance Summary

		Base Case	Air Preheat Case
Ethane feed	t/h	51.9	51.9
Ambient conditions	-	15°C, 85% R.H.	15°C, 85% R.H.
Air temp. to burners	°C	15	129
Stack temperature	°C	169	64
Excess air	%	10	5
Absorbed heat	MW	116.8	110.6
Overall efficiency	%	91.7	96.5
Fuel flow	t/h	8.2	7.3
Fired heat	MW	127.3	114.5
VHP steam production	t/h	97.0	88.75
NOx concentration		Base	Base + 2%

Table 6
Economics of Ethane Furnace with Air Preheating

		Air Preheat Case (EU location)	Air Preheat Case (US location)
Savings in fuel firing	million €/yr	3.01	1.83
Cost of lost VHP steam capacity	million €/yr	1,64	1.00
Cost of additional electricity	million €/yr	0.10	0.06
Net OPEX saving	million €/yr	1.27	0.77
Payback	Years	1.5	3

Fig 1
Naphtha cracking furnace

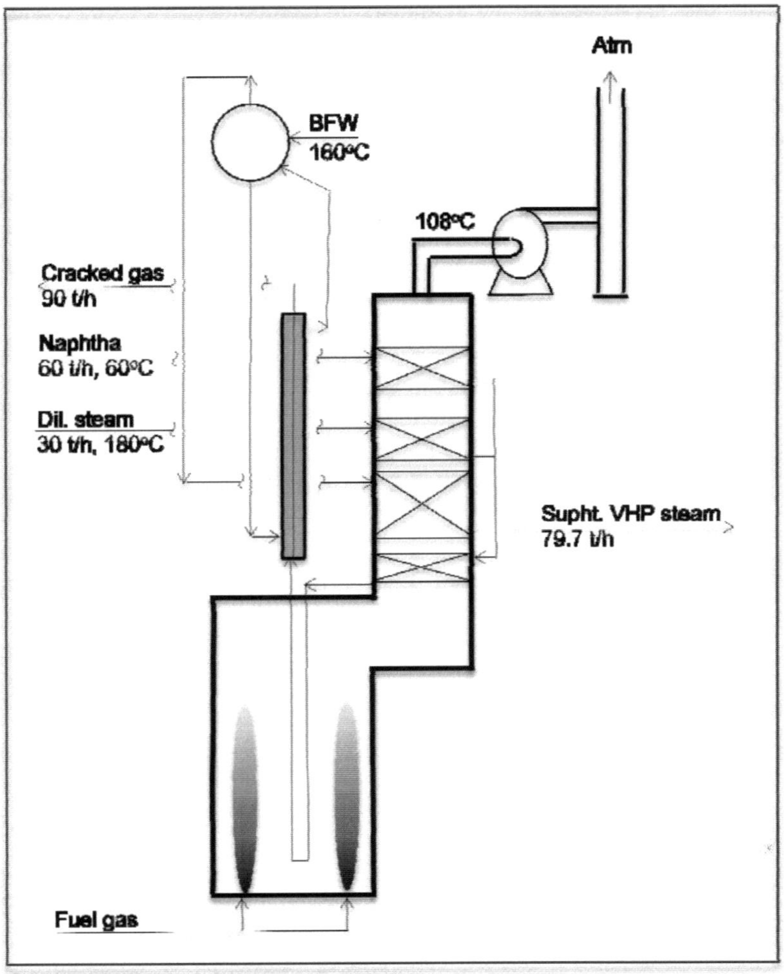

Fig 2
Ethane cracking furnace

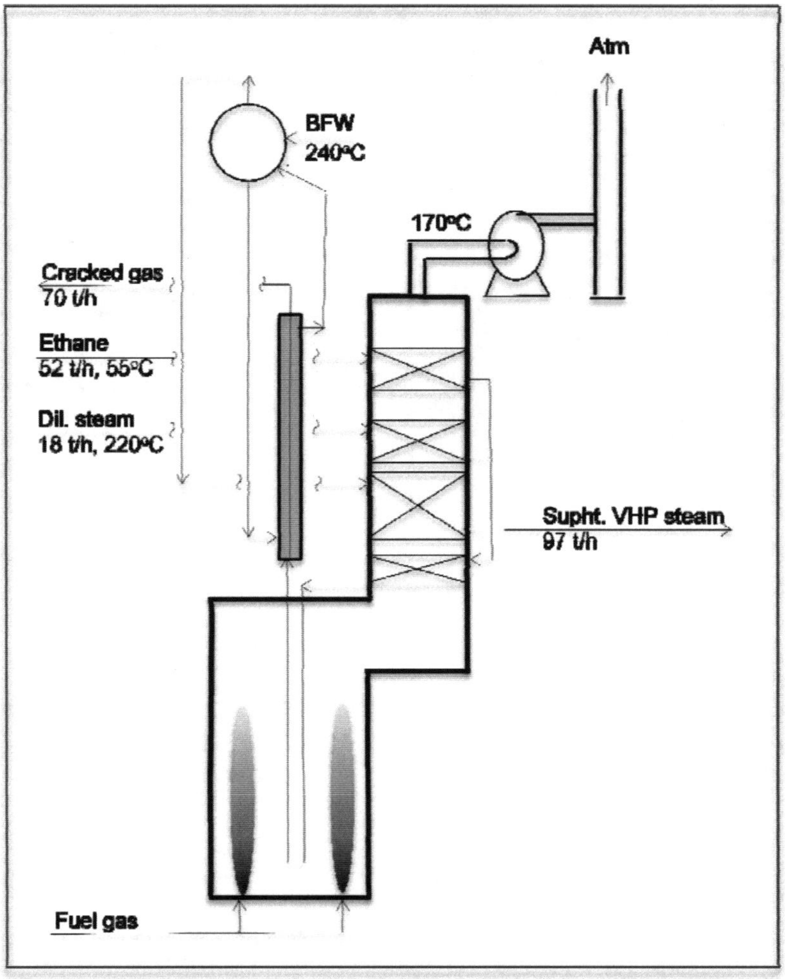

Fig 3

Naphtha cracking furnace – Air Preheater option

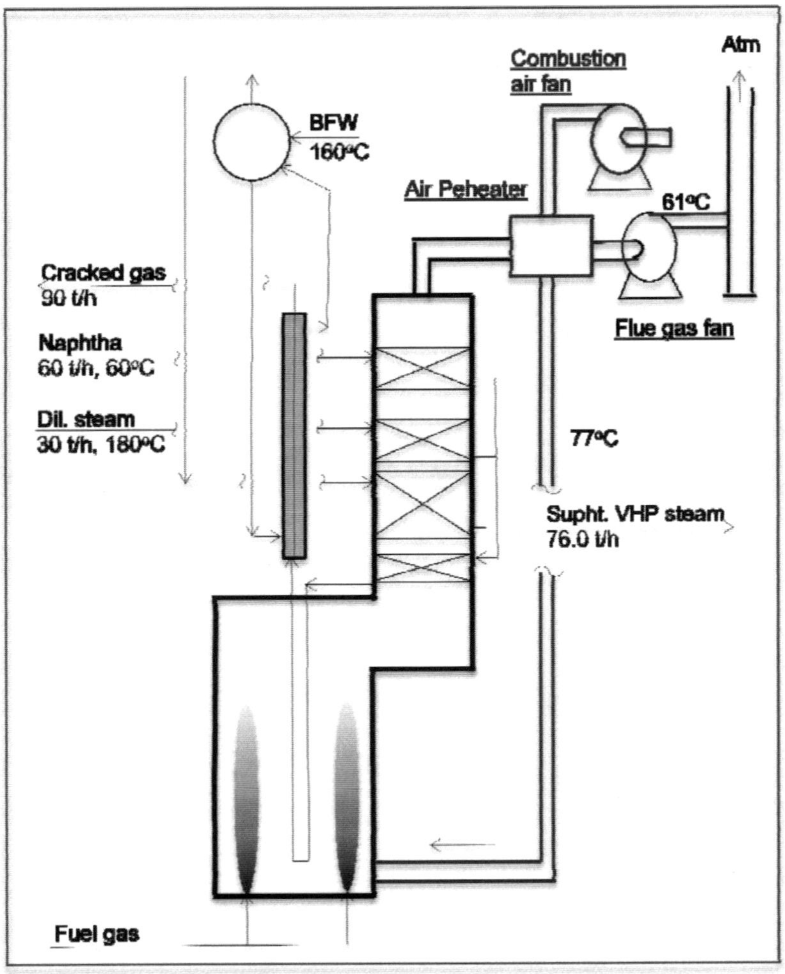

Fig 4

Ethane cracking furnace – Feed Preheater option

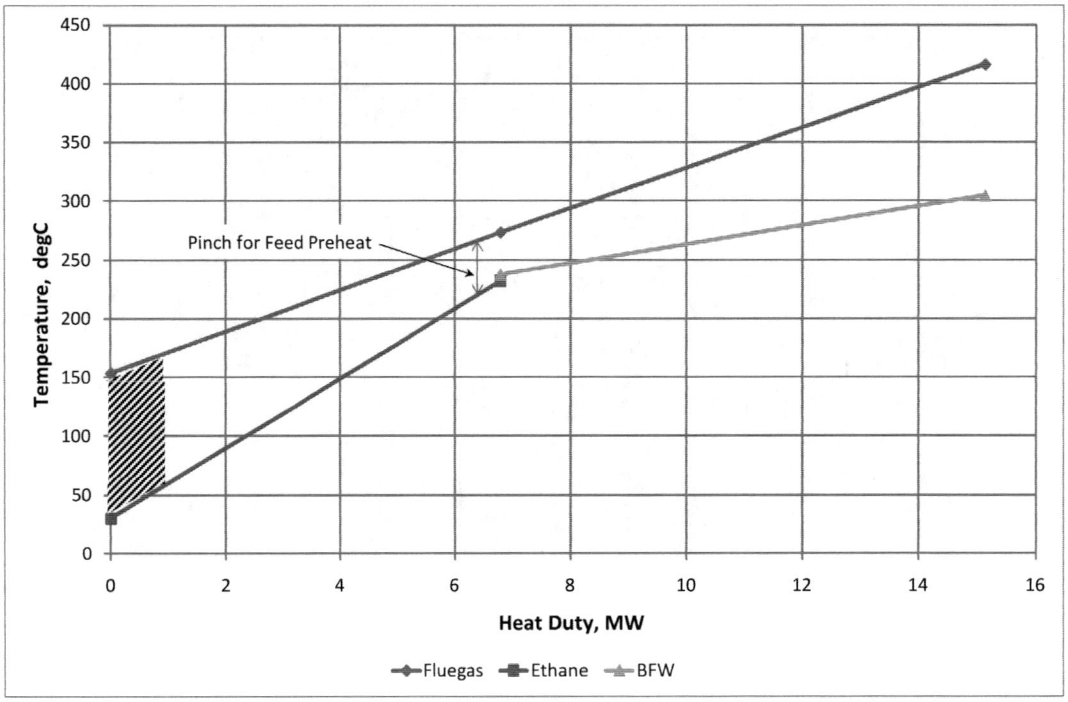

Modernize Your Control Functionality
Easy Steps to Start Process Optimization

13 AIChE
Spring Meeting, San Antonio, TX

Presenters

- Laurie Ben
- James Beall

EMERSON.
Process Management

Agenda

- Introduction
- Examples – Best Practices & Lessons Learned
  - Split Range Duty
  - Smith Predictor
  - Gain Scheduling/Adaptive Control
  - Signal Characterization
- Summary

Introduction

- Modernizing Control System – Decision Done Check!
- 'Do I' or 'Don't I' take advantage of new technology
  - Implemented older cumbersome techniques?
  - Complex custom strategies that handle automation challenges?
  - Hard to implement & maintain?
- Standard features that are field proven, practical, and easy to implement
  - Easier to implement & maintain
  - More efficient and effective control strategies
  - Achieve improvements in process performance
  - Lays the foundation for optimization

Agenda

- Introduction
- Examples – Best Practices & Lessons Learned
  - Split Range Duty
    - Control Challenge
    - Legacy Solution
    - Newer Technology Solution
    - Summary
  - Smith Predictor
  - Gain Scheduling/Adaptive Control
  - Signal Characterization
- Summary

Split Range Duty – Control Challenge

- Use steam and hot oil as energy source for distillation column for temperature controller.
- Hot oil is "free energy" and should be maximized
- Control Scenarios
  - Maximize hot oil and supplement with steam
  - Base load steam (if meter bad) and supplement with hot oil
  - Base load hot oil flow (if bad meter) and supplement with steam
- Constraints
  - Minimum and maximum hot oil valve position and flow rate
  - Minimum steam flow rate

Split Range Duty – Legacy Solution

- Old legacy system implemented this feature using a velocity PID with incremental output and a “micro expert rules system”

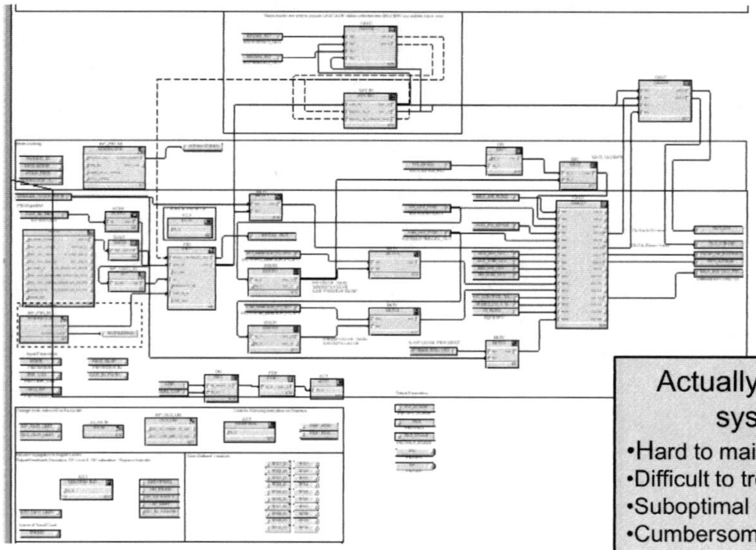

Actually Duplicated in New system, but still…

- Hard to maintain and modify
- Difficult to troubleshoot
- Suboptimal performance
- Cumbersome/Kludge/Ugly

Split Range Duty – New Technology Solution

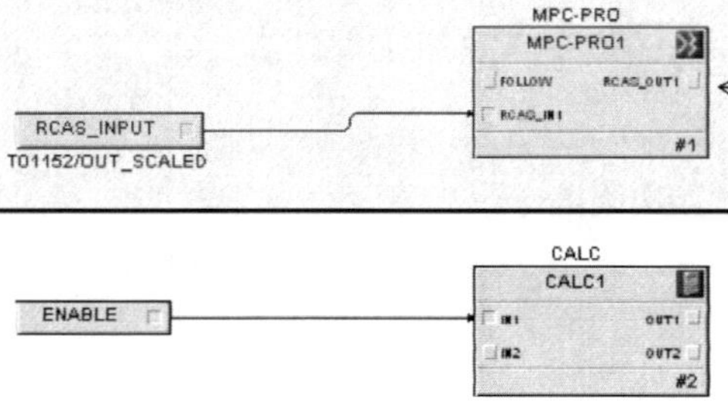

Embedded MPC Function Block provides all control functionality! EASY!!!

Simple calculations to calculate limits

Split Range Duty – New Technology Solution

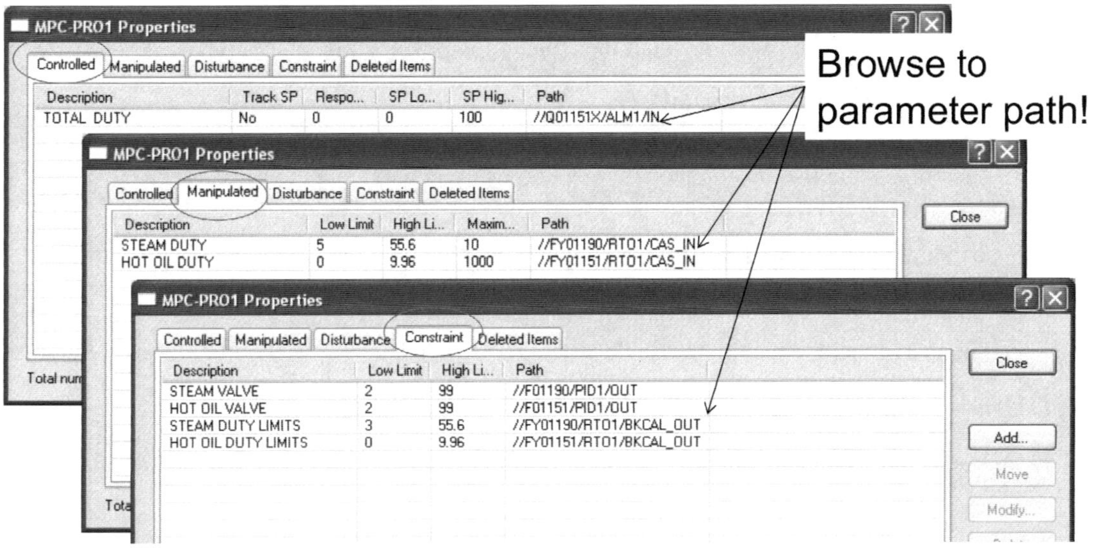

Split Range Duty – New Technology Solution

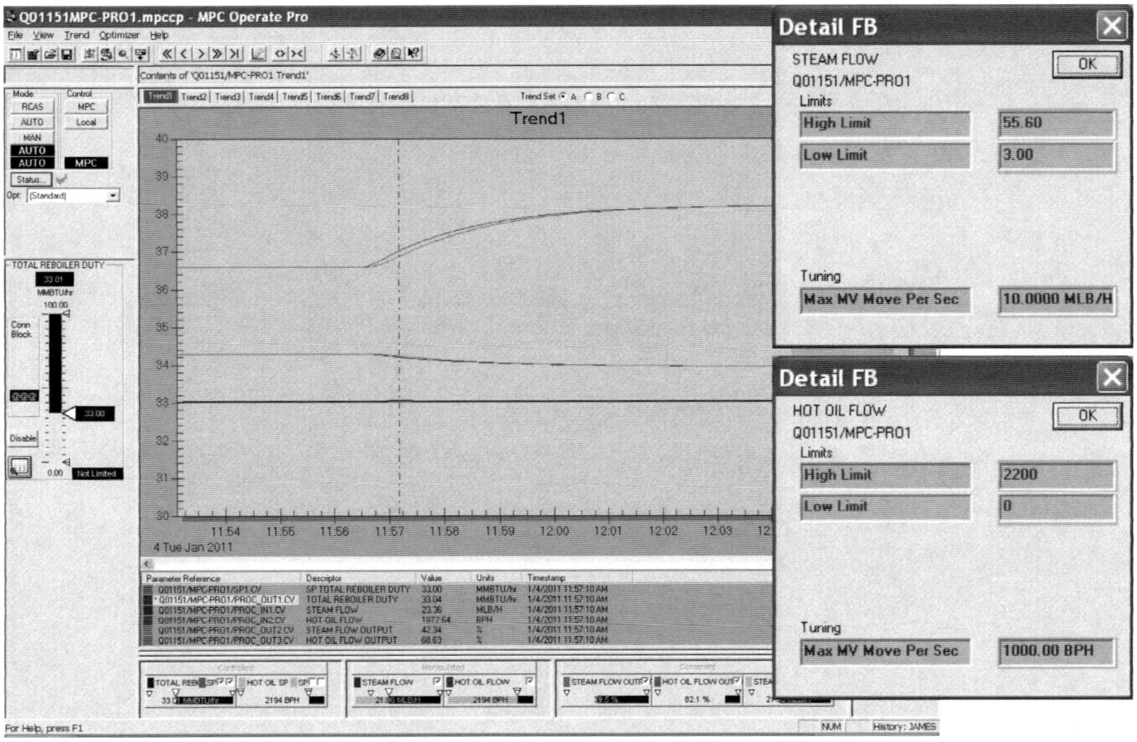

Summary Split Range Duty Control

Embedded MPC (New Technology) vs Legacy Solution

- Better performing solution
  - Bumpless transfers and constraint control
- Easier to Maintain
  - Change limits and settings
  - Simpler implementation
  - Easy to understand, train others, document
- Easily expands to handle future constraints
- Reduce energy costs!

Agenda

- Introduction
- Examples – Best Practices & Lessons Learned
  - Split Range Duty
  - Smith Predictor
    - Control Challenge
    - Legacy Solution
    - Newer Technology Solution
    - Summary
  - Gain Scheduling/Adaptive Control
  - Signal Characterization
- Summary

Smith-Predictor PID – Control Challenge

- Control of processes that are "dead time dominant" (dead time > time constant)
- Typically custom configured
  - some legacy control systems have preconfigured options, e.g. PROVOX
- Algorithm uses a process model
  - (gain, dead time and time constant)
- Challenge: as process dynamics change from model used, performance degrades rapidly

Smith-Predictor PID – Legacy Solution

- Traditional Smith-Predictor

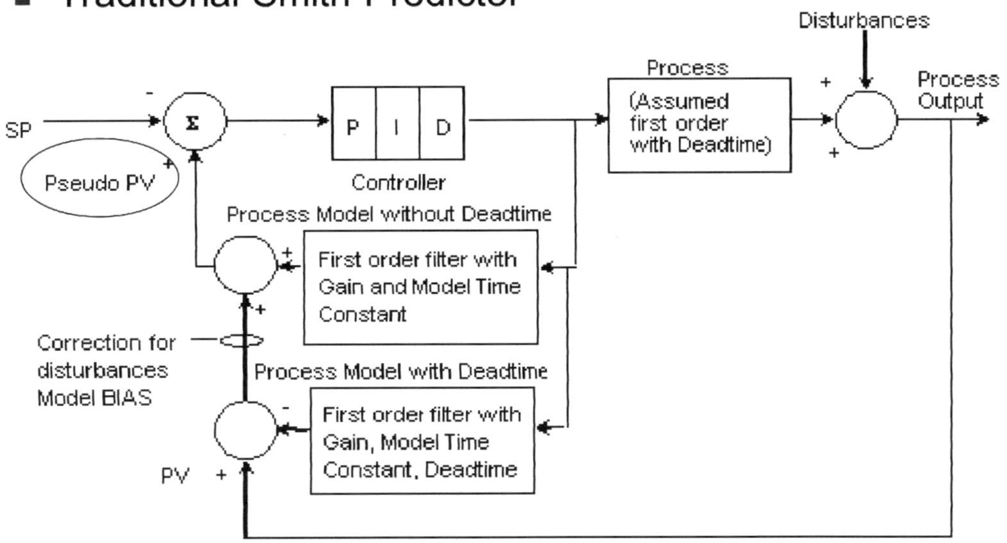

•High maintenance when process dynamics change
•Performance degrades rapidly as process dynamics change from configured model

Smith-Predictor PID – New Technology Solution #1

Easy to implement the Smith-Predictor but…

- High Maintenance when process dynamics change
- Performance degrades rapidly as process dynamics change from configured model

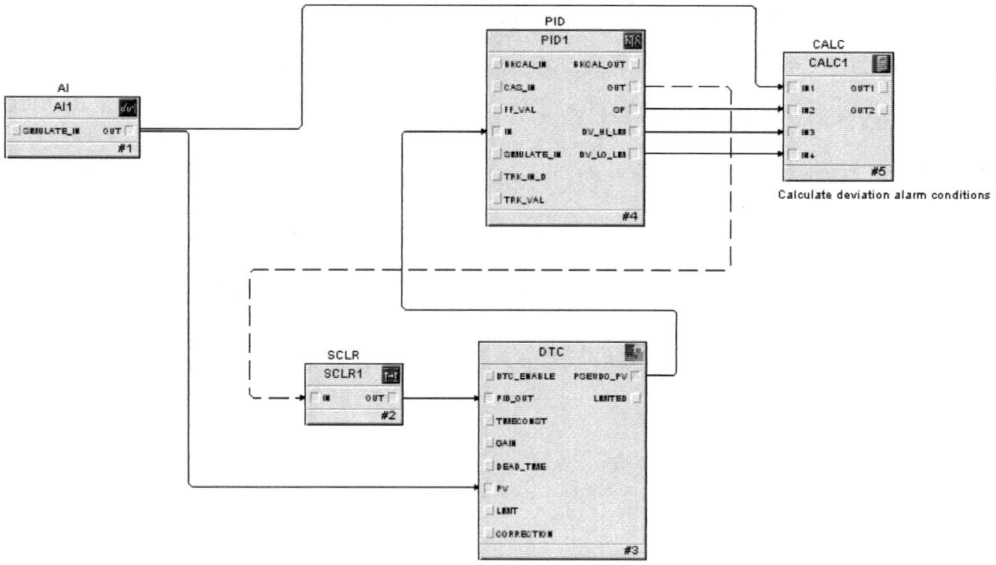

Smith-Predictor PID – New Technology Solution #2

- Embedded MPC Solution

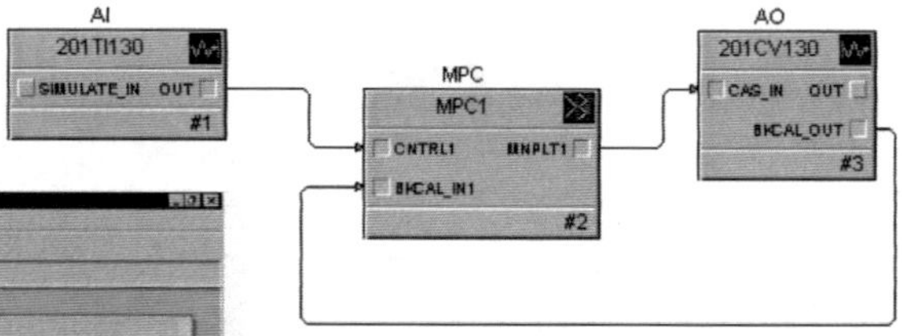

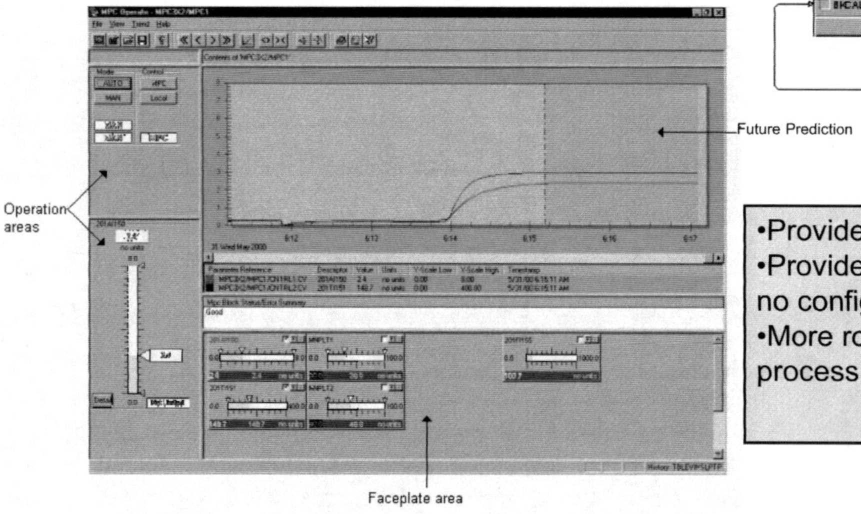

- •Provides Future Prediction
- •Provides MPC Faceplate – no configuration needed
- •More robust under changing process dynamics

Smith-Predictor PID – New Technology Solution #2

- Embedded MPC Solution

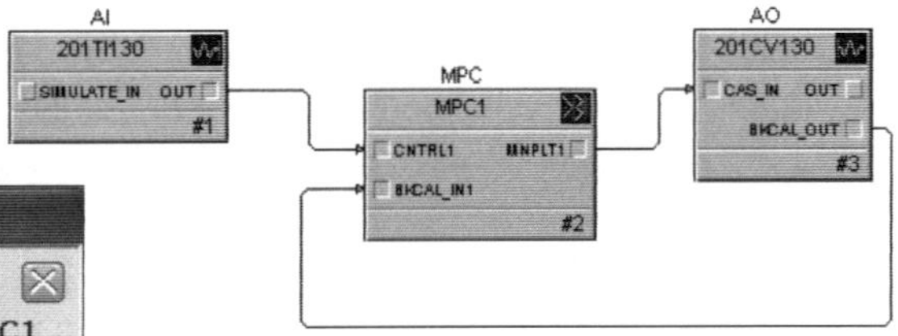

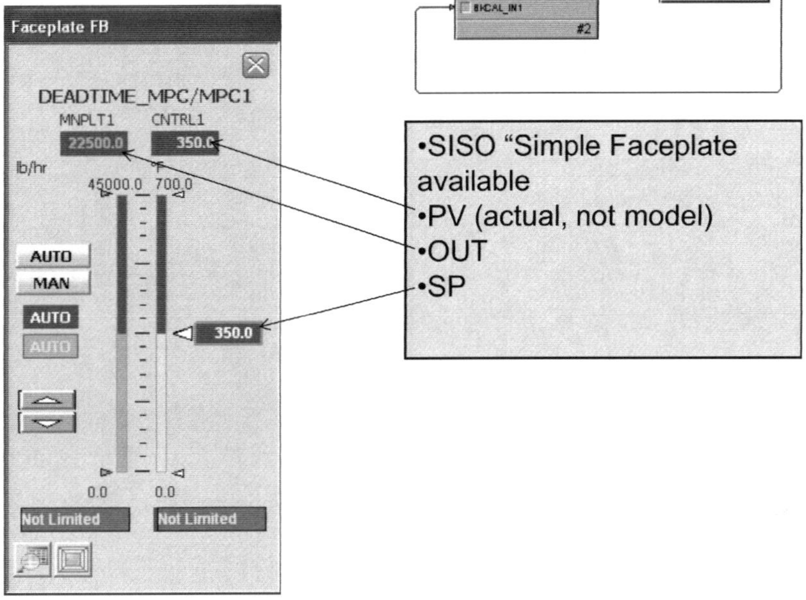

Summary Smith-Predictor PID

New Technology Solutions

- Standard Smith Predictor (#1) and Embedded MPC (#2)
- Both easy to implement and understand!

New Technology Solution #2 (Embedded MPC)

- More robust solution under a changing process model
- Provides visual display of the "future" PV
- Save money through better control performance

Agenda

- Introduction
- Examples – Best Practices & Lessons Learned
  - Split Range Duty
  - Smith Predictor
  - Gain Scheduling/Adaptive Control
    - Control Challenge
    - Legacy Solution
    - New Technology Solution
    - Summary
  - Signal Characterization
- Summary

Gain Scheduling – Control Challenge

- Legacy systems have limited capability to modify tuning based on process conditions
- Others had add-on application software, e.g. PROVOX "Gain Scheduler" system provided 3 sets of Gain, Reset, and Rate applied in 3 separate regions of a "state variable"
  - State variable = PV, OUT or "other"
  - Transition between tuning sets interpolated over "dead band"
- Beware: some suspend loop execution when tuning parameters change, so don't change the tuning every scan

Gain Scheduling – Legacy Solution

- Custom code
- Multiple points, registers, displays
- Restricted (slower) update rate

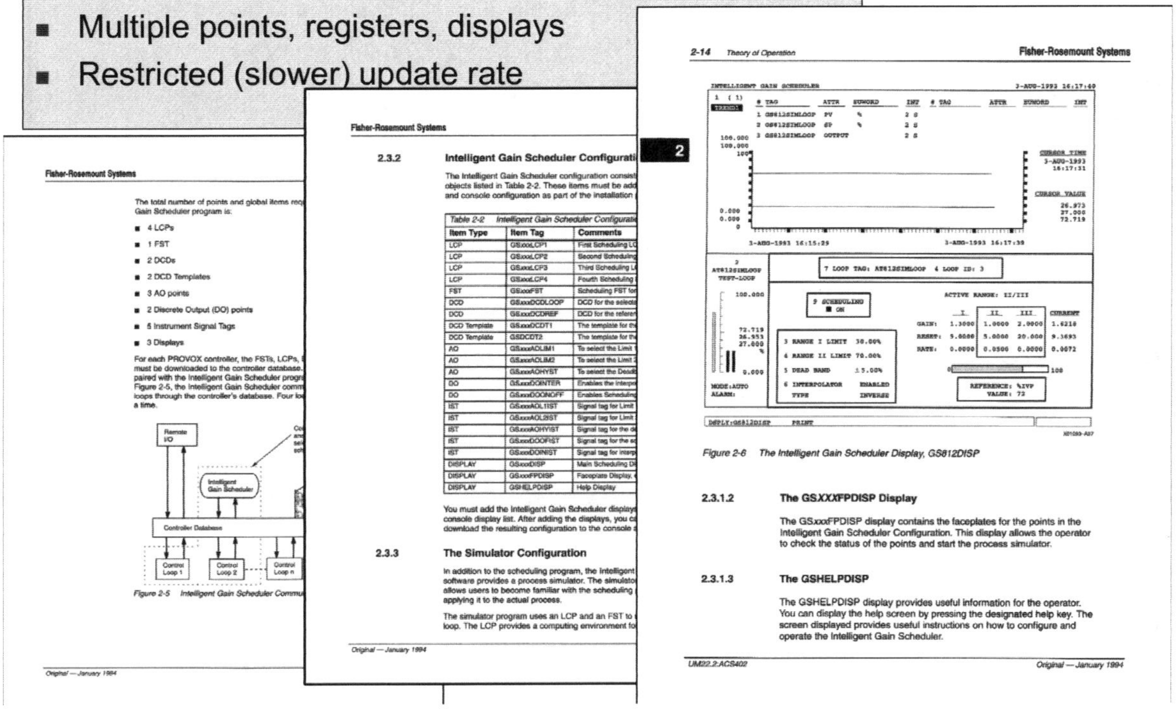

Gain Scheduling – New Technology Solution #1

Standard Modules

- Standard configuration
- Preconfigured faceplate & detailed display
- Easy to implement
- Tuning updates at loop execution

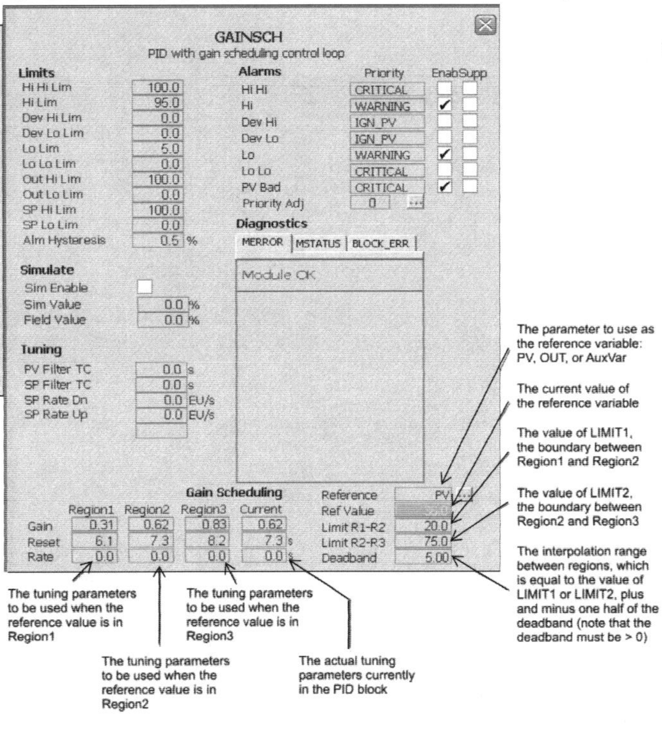

Gain Scheduling – New Technology Solution #2

New Technology "Adaptive Control"

- "Learns" process models from normal operating moves
- Allows up to 5 tuning regions of the "state variable"
- Partial Adaptive Control
  - Use "approved model" only & selected tuning rule in each region
  - Lambda PI or IMC PID; "Closed loop time constant" or "Lambda Factor".
- Optional Full Adaptive Control
  - continuous "Learning" of process model in each region
  - updates tuning while in each region per selected tuning rule
  - set allowable deviation (%) from approved model in each region

New Technology – Adaptive Control

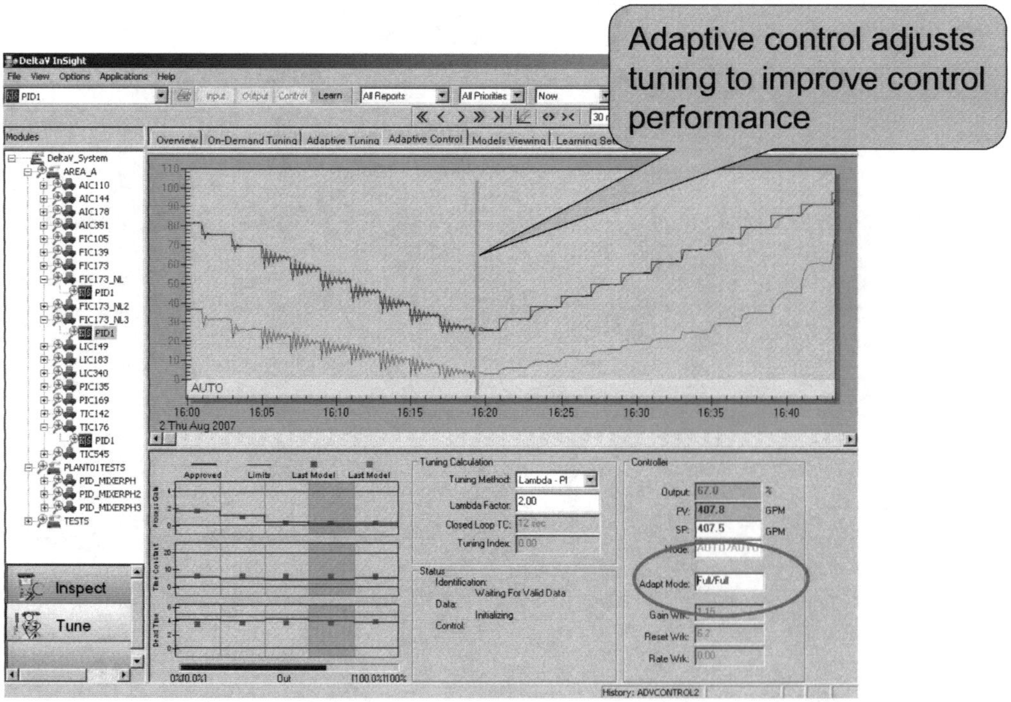

Summary Gain Scheduling

New Technology Solution(s) vs. Legacy Solution

- Gain Scheduling easy to implement and understand!
  - Standard configuration
  - Preconfigured faceplate & detailed display
  - Tuning updates at loop execution
- Adaptive Control provides additional benefit and ease of implementation
  - Learns models during normal operation – NO STEP TESTs!
  - More Regions – Up to 5
  - Automatic tuning updates when process model changes
- Improved control performance saves money!

Summary Gain Scheduling

New Technology Solution(s) vs. Legacy Solution

- Gain Scheduling easy to implement and understand!
  - Standard configuration
  - Preconfigured faceplate & detailed display
  - Tuning updates at loop execution
- Adaptive Control provides additional benefit and ease of implementation
  - Learns models during normal operation – NO STEP TESTs!
  - More Regions – Up to 5
  - Automatic tuning updates when process model changes
- Improved control performance saves money!

Agenda

- Introduction
- Examples – Best Practices & Lessons Learned
  - Split Range Duty
  - Smith Predictor
  - Gain Scheduling/Adaptive Control
  - Signal Characterization
    - Control Challenge
    - Legacy Solution
    - New Technology Solution
    - Summary
- Summary

Signal Characterization - Control Challenge

- Input Characterization (AI)
  - Combustion control – characterize fuel and air
  - Level – characterize pv (e.g. level to volume)
- Output characterization (AO)
  - modify installed flow characteristics of valves, dampers or other final control elements
  - Use to modify loop gain, e.g. output path to slave loop
- Legacy System Challenges
  - Lack of resolution (few points)
  - External black box characterizers may shift, lose documentation
  - Potential operator/maintenance confusion on signal going to final control element

Signal Characterization – Legacy Solution

- Legacy system using external analog characterizer

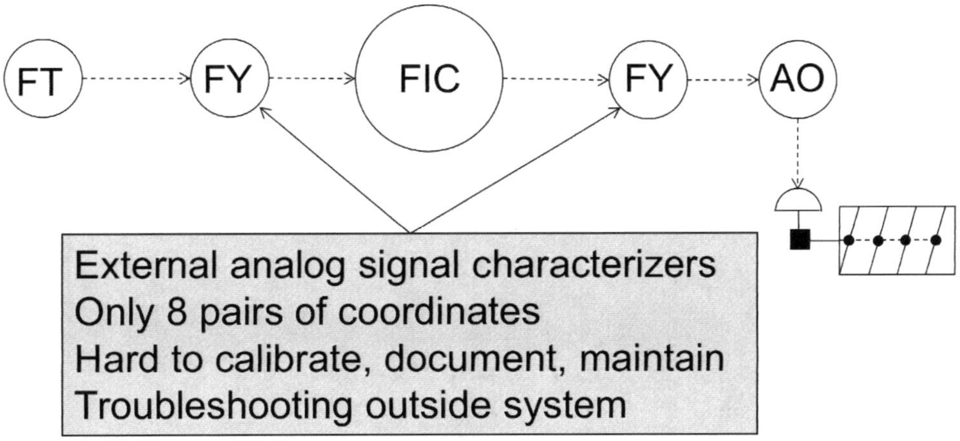

Signal Characterization – New Technology Solution #1

- Standard Module implementation provides 21 point signal characterizer function block (SGCR)

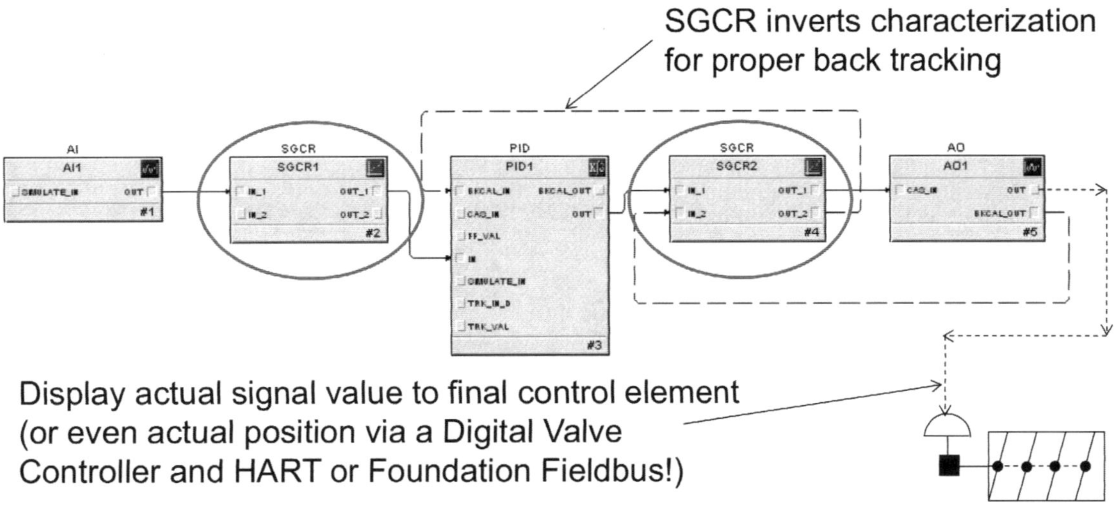

Signal Characterization – New Technology Solution #2

- Adaptive Control
  - Adjust tuning to compensate for final control element characteristics

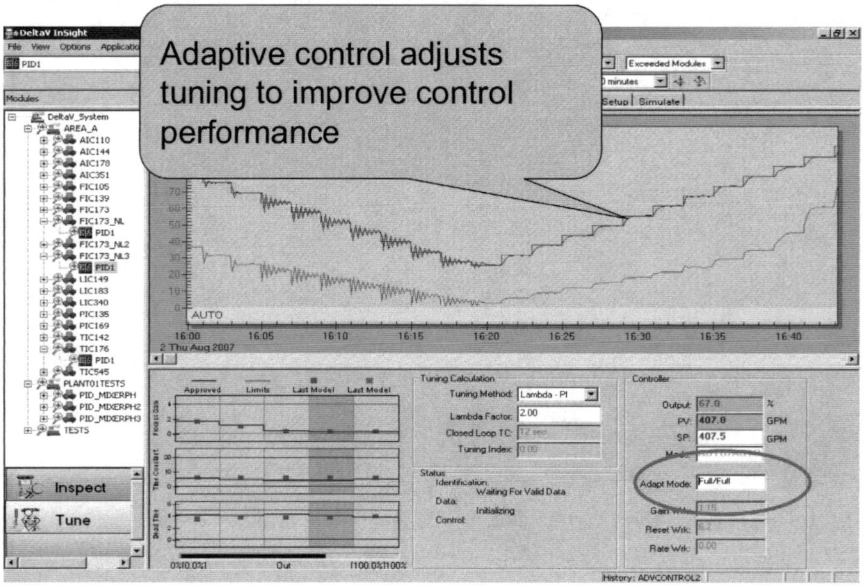

Summary Signal Characterization

- Newer Technology Signal Characterization function block
  - Better resolution (21 pairs of coordinates)
  - Self documenting (exists in control system)
  - No calibration required vs. 'external black box'
  - Provides "inverse" characterization for analog output characterizations
  - Reduce confusion: display signal being sent to final control element
- Applying Adaptive Tuning may provide a superior solution for AO characterization of final control element

Agenda

- Introduction
- Examples – Best Practices & Lessons Learned
  - Split Range Duty
  - Smith Predictor
  - Gain Scheduling/Adaptive Control
  - Controller Characterization
- Summary

Summary

- Practical application of control features that deal with common challenges
- Highlighted best practices & lessons learned
  - Split Range Duty
  - Smith Predictor
  - Gain Scheduling/Adaptive Control
  - Signal Characterization
- Do look at new functionality available when modernizing
  - Allows jump start to process optimization
  - Leads to improved process performance
  - Easier to implement and maintain over the life investment

Contacts & Questions?

James.Beall@Emerson.com
903-235-7935

Laurie.Ben@Emerson.com
512-834-7237

Paralleled Double-Effect Distillation: Simulative Case Studies

*Pei Wang, Leilei Dai, Ming Shi , Fan Wang, Hong Gao, Jinsheng Sun**
School of Chemical Engineering and Technology, Tianjin University, No. 92 Wei Jin Road, Tianjin 300072, P. R. China

Abstract

The increasing scarcity and growing demand for energy have led to an increasingly attention to the energy-efficient methods. Double-effect distillation is a common method to achieve remarkable energy conservation, which has three kinds of typical configurations including the feed splitting (FS), the light split/forward (LSF) and the light split/reverse (LSR). In spite of a good pathway for energy saving, better options are expected to cut down energy expenditure by a bigger margin than double-effect distillation does.

In this paper, case studies of paralleled or combined double-effect distillation schemes are proposed on the basis of current double-effect distillation. In order to verify the feasibility of this method, the improved methanol distillation scheme and organosilicon distillation scheme are optimized under this principle as examples. Exergy losses and the energy consumptions are measured as a tool to compare the two prevalent schemes with their corresponding optimized counterparts, respectively. The results of the total exergy loss and the energy consumption of these schemes show that the optimized schemes can cut the corner of the exergy loss by 22.8% and 40.41% and the energy consumption by 15.61% and 44.28%, respectively, in comparison to their popular cousins.

Keywords: double-effect distillation, paralleled double-effect distillation, methanol distillation, organosilicon monomer distillation, distillation process simulation, energy saving, exergy analysis

1. Introduction

Distillation, in chemical process and related industries, is by far the most widely applied for the separation of liquid mixture. In spite of its well-known benefit and widespread application, the major drawback is its high energy consumption. In classical distillation process, heat is transferred to the reboiler of distillation column. Along with the distillation column to separate the mixture, a large number of heat are directly discharge into the surrounding environment and cooling water. Araujo AB [1] has done some research to achieve that distillation has a relatively low thermodynamic efficiency and it requires the input of high quality energy in the reboiler to push the separation of liquid mixture. Therefore, the focus of energy conservation is to consider how to recover or reuse the discharged heat, or how to reduce the quantity of transferred heat to the column. In the early 20th century, many scholars have devoted to the study of saving energy in distillation, and now most of correlative techniques are widely used in distillation process [2,3]. These energy saving methods involve multi-effect distillation, heat pump distillation, heat integrated distillation and etc [3-8]. But without extra rotating machines by electricity or other power producers, multi-effect distillation has been more favored in research and application [9].

Many researches confirmed multi-effect distillation to be one of the effective methods to reduce the energy consumption [10-12]. The energy saving effect of those schemes used multi-effect distillation is associated with the number of the effects and the characteristics of separation system. Generally speaking, those schemes can offer energy savings around 30% with respect to the original or classical multicomponent distillation scheme [23, 27, 28]. Considering the interaction between investment of plants and energy saving effect, the double-effect distillation has been the most commonly used in not only research but also field application.

The use of the double-effect distillation technology for optimizing the conventional distillation schemes can cut a large amount of the energy requirement in comparison with the correspondingly prevalent schemes [13]. Although the system is very difficult to control in the sense that it is nonlinear, multivariable and interacting [14]. More and more published studies [15-19] have been focused on further questing the way of energy saving on the basis of the double-effect distillation schemes. However, the overhead vapor of one high pressured column only can integrate with the reboiler of one low pressured column in those published studies. In view of this, in this paper, the paralleled or combined double-effect distillation is proposed to broaden the thought for the application of the double-effect distillation, in which the overhead vapor from one of the high pressured column is divided into two or more streams to successively integrate with the reboiler of the other columns [9].

2. *Theory*

2.1 Double-effect or heat integrated distillation system

A double-effect distillation system is made up of two adjacent distillation columns running in different pressure, which can lead to a temperature difference between the top of the high pressured column and the bottom of the low pressured column. Through the thermodynamic connection, the reboiler of the low pressure column plays a role of the condenser for the high pressure column. According to the direction of the mass flow and the heat integration, the double-effect distillation can fall into three kinds of configuration: the feed splitting (FS), the light split/forward (LSF) and the light split/reverse (LSR), as showed in Figure. 1.

One of the configurations with a high thermodynamic efficiency is the feed splitting (FS), in which the feed is divided roughly equal between the two columns running in different pressure, and the products can be obtained from both of the top and bottom of the two columns, as illustrated in figure. 1 (a). The overhead vapor form the high pressured column is used to heat the reboiler from the low pressured column. There is only energy exchange between the two columns, so the heat duties of the feed splitting (FS) are approximately equal to the heat duty for single column with the same conditions divided by the number of effect.

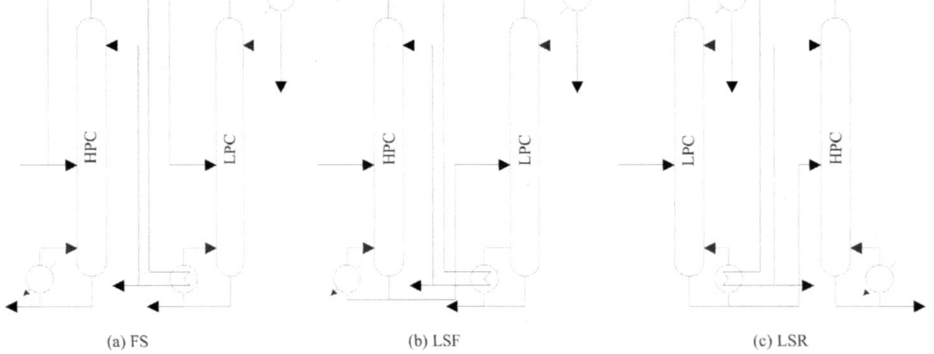

Fig. 1. Three configurations of double-effect distillation

Figure. 1 (b) and (c) respectively shows the configuration of the light split/forward heat integration (LSF) and the light split/reverse heat integration (LSR). Different from the feed splitting (FS), the total feed stream was sent to one column in the other two configurations. In the light split/forward (LSF), the total feed is pumped to the high-pressure column (HPC) and the bottom stream of HPC flows into the low-pressure column (LPC). In this way, all of the heavy components are from LPC and the light components from the top both of the two columns. Figure. 1 (b) indicates the direction of mass flow and heat integration is the same. The last configuration is the light split/reverse (LSR), as shown in figure. 1 (c), all of the feed is sent to the low-pressure column (LPC). It can remove the two kinds of light components from the top both of the two columns and one kind of heavy components from the bottom of HPC. Moreover, the direction of heat integration is in the reverse with the mass flow. These two kinds of configurations have not only energy exchange but also substantial exchange, therefore, when the operating parameters of one column are changed, the other column will also change.

Theoretically, the three kinds of configurations are empirically 30% less energy consumption than the single column distillation. As for choosing what kind of the configurations of the double-effect distillation, it depends on the properties of the multi-component mixture, the pressure of saturated steam, the temperature of cooling water, structure parameters of existing distillation column and so on.

2.2 Exergy analysis of the distillation process

According to the First Law of Thermodynamics, also called the Law of Energetic Balances, no matter what happens, the total amount of energy has not yet changed. However, when the various types of energy generate the mutual conversion, the quality of the energy is always decreased. The separation process in distillation is irreversible. In order to separate the mixture in distillation column, it needs the driving force which must consume a certain amount of external work, by this means, the decline of the quality of energy is inevitable. Therefore, we should choose a reasonable driving force in distillation process to avoid unnecessary loss of the quality of energy. It defined the concept of exergy (E_x) to compare the quality of the different types of energy. Exergy, which is a measure for evaluating the extent of available energy or the quality of energy, is based on the Second Law of Thermodynamics. Exergy analysis reveals the essence of the energy, and can be used as an important tool to evaluate the effect of energy saving. And it usually combines with process optimization studies, and as its guidance, to identify cost saving opportunities and amend the existing scheme.

Szargut [29] has summarized a comprehensive definition of the exergy, that is: 'Exergy is the amount of work obtainable when some matter is brought to a state thermodynamic equilibrium with the common components of the natural surroundings by means of reversible processes, involving interaction only with the above mentioned components of nature'. The state of one stream recorded as 1, according to the definition of exergy, the exergy of the stream can be expressed as:

$$E_x = (H_1 - H_0) - T_0(S_1 - S_0) \quad (2)$$

where, H_1 and S_1 is the enthalpy and the entropy of the stream 1, and T_0 is the temperature of the surrounding and H_0 and S_0 represent the enthalpy and the entropy of the surrounding.

The state of another stream recorded as 2, when the system stream change from state 1 (T_1, P_1) to state 2 (T_2, P_2), the variation of exergy (ΔE_x) is given by

$$\Delta E_x = E_{x,1} - E_{x,2} = \Delta H - T_0 \Delta S \quad (3)$$

In the conventional distillation process, it must input the heat or energy to drive the occurrence of separation process. The bottom of the column input the heat (Q_R) at this temperature (T_R), and the top of the column was sent to the heat (Q_C) at this temperature (T_C). The net work consumption (W) in distillation is

$$W = Q_R(1 - \frac{T_0}{T_R}) - Q_C(1 - \frac{T_0}{T_C}) \quad (4)$$

When the enthalpy difference between the product and the feed is very small, $Q_R=Q_C=Q$ can be approximately gotten, then Equation (4) is rewritten to be:

$$W = QT_0(\frac{1}{T_C} - \frac{1}{T_R}) \quad (5)$$

Then, the total exergy loss (Ex_{loss}) of the distillation column can be calculated by

$$Ex_{Loss} = W - \Delta E_x \quad (6)$$

3. *Prototype case: the prevalent distillation schemes*

3.1 The improved four-column methanol distillation scheme

In the methanol industry, it has gradually appeared two-column, three-column, four-column and five-column distillation scheme [20-23]. With the increase of the number of distillation column, the corresponding distillation scheme can not only increase the efficiency of the process, but also cut the energy consumption. At present, the four-column scheme has a dominant status to produce the purified methanol. However, the improved four-column distillation scheme can cut some more energy requirement with respect to the traditional four-column distillation scheme. The improved four-column distillation scheme is seen as the comparative reference in the paper.

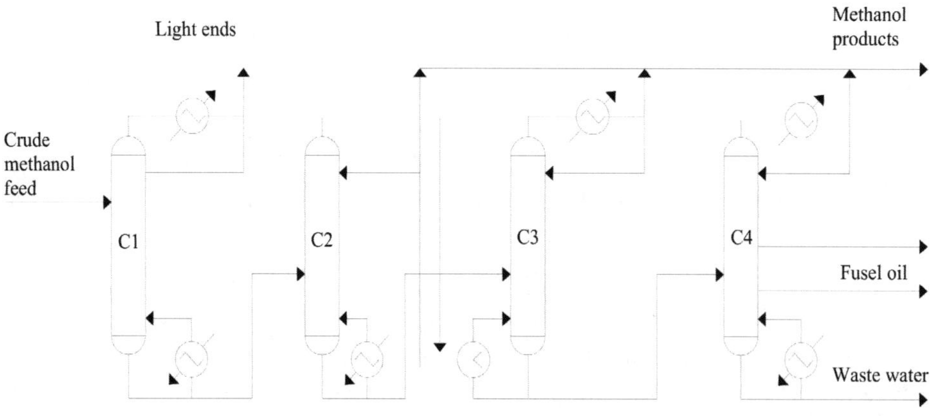

Fig. 2. The improved four-column double-effect methanol distillation scheme (C1, light ends column; C2, pressured column; C3, atmospheric column; C4, water column)

In the improved four-column distillation scheme, as shown in figure. 2 [24], the crude methanol flows into the light ends column (C1) in 87.5 ℃ for removing the light ends and the remnants of dissolved gas from the top of C1. Besides, in order to raise the methanol recovery rates, the light ends column (C1) designs a two-stage condenser system. The crude methanol was distillated successively from the pressured column (C2) and the atmospheric column (C3), and most of the methanol products can be received at the top both of C2 and C3. The bottom stream of C1, C2 and C3 is pumped respectively as the feed into C2, C3 and the water column (C4). Residual methanol, fusel oil and waste water are extracted from the corresponding position of the top, side stream and the bottom of the water column (C4).

In the improved scheme, the operation pressure of C2 is 0.9MPa, and the other three columns run under the atmospheric pressure. Due to the different of operating pressure, the overhead vapor of C2 can undertake thermal coupling connection with the reboiler of C3.

3.2 The current organosilicon monomer distillation scheme

According to the survey, there are several kinds of organosilicon distillation schemes. However, the 9-column organosilicon distillation scheme [9], in China, has an important role in the organosilicon industrial, as illustrated in Figure. 3, it is used as the base case.

Figure. 3 shows that the crude monomer is delivered to the heavies removal column (C1') and the bottom stream of C1' has a lot of the unsteady substances of high boiling points. These unsteady substances of high boiling points are pumped to the high boiling point column (C9') as the feed, and the products from the top of C9' continue to serve as the crude monomer by circumfluence. By this way, it can accomplish the separation between the unsteady substances of high boiling points and the useful substances at a relatively low temperature in the high boiling point column (C9') and avoid the unsteady substances of high boiling points from thermal cracking to generate many unknown substances which would greatly affect the subsequent separation of product. The low removal column (C2') provides trimethylchlorosilane (Me3), methyldichlorosilane (MeH), and other light components for lights removal column (C3'), MeH column (C4'), azeotrope removal column (C5') and Me3 column (C6'), which can finally achieve two products (MeH and Me3) and three wastes. It also provides methyltrichlorosilane (Me1) and dimethylchlorosilane (Me2) and other heavy components for Me1 column (C7') and Me2 column (C8'), which can finally get two products (Me1 and Me2).

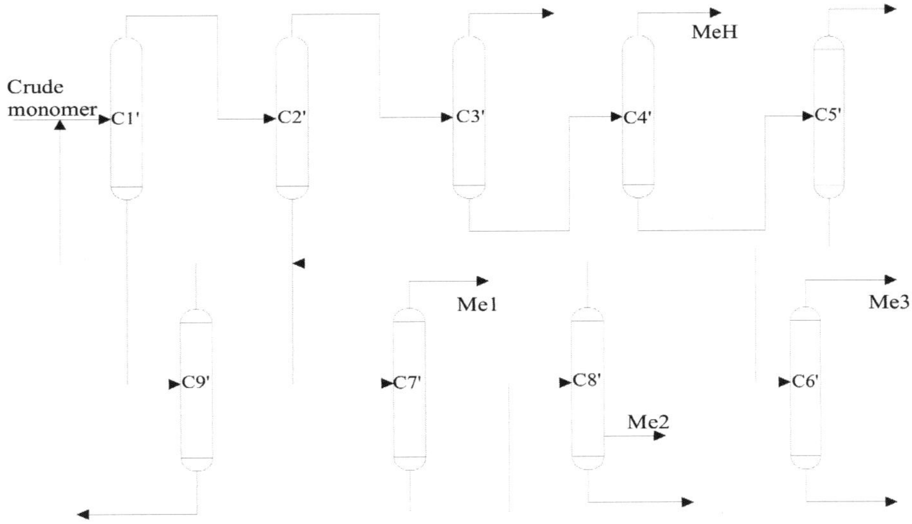

Fig. 3. The prevailing 9-column distillation scheme (C1', heavies removal column; C2', low removal column; C3', lights removal column; C4', MeH column; C5', azeotrope removal column; C6', Me3 column; C7', Me1 column; C8', Me2 column; C9', high boiling point column)

In this scheme, the operating pressure of all nine columns is relatively low, and the highest operating pressure is 260 kPa in the lights removal column (C3'). Although it has a different operating pressure between C3' and the other columns, but the temperature difference is not enough to make them use double-effect or heat integrated distillation.

4. Simulation and analysis both of the prevalent schemes and the optimized schemes

4.1 The novel five-column heat integrated methanol distillation scheme

Although the energy saving of the improved four-column methanol distillation scheme can reach 10% compared with the traditional four-column scheme, it still exist several shortages, which will be illustrated by the results of computer simulation to the improved four-column methanol distillation scheme. The flow rate of the crude methanol is 239230 kg/hr, and the temperature is 87.5 ℃, and the composition of the crude methanol is exhibited in table 1. The inlet and outlet temperature of the cooling water are 30 ℃ and 40 ℃.The methanol content is specified as 99.99 wt% in the methanol products and 10ppm in the wastewater of the water column. Table 2 shows the operating parameters of each column in the improved four-column scheme.

Table 1. Crude methanol feed data

Component	Mass fraction%	Component	Mass fraction%
H_2	0.0100	Pentanol	0.0338
N_2	0.0710	Formic acid	0.0030
Methane	0.0100	Acetic acid	0.0030
CO	0.0360	Methyl formate	0.0092
CO_2	0.6640	Acetone	0.0193
Methanol	95.0000	Methyl ethyl ketone	0.0030
Dimethyl ether	0.0309	Diisopropylether	0.0056
Trimethylamine	0.0010	Water	3.7100
Ethanol	0.0900	i-Pentane	0.0562
i-Propanol	0.0024	n-Pentane	0.0169
Propanol	0.0264	n-Hexane	0.0487
i-Butanol	0.0338	n-Heptane	0.0009
Butanol	0.0185	n-Octane	0.0050
3-Pentanol	0.1407	n-Nonane	0.0050

Table 2. The operating parameters of improved four-column scheme

The operating parameters	C1	C2	C3	C4
Top pressure (kPa)	150	900	110	110
Top temperature (℃)	81.0	132.6	67.7	67.7
Bottom temperature (℃)	83.8	139.1	96.5	104.5
Reflux ratio	0.58	2.44	1.68	11.57
Number of theoretical stage	33	52	58	61

The results of simulation indicate that the hot utility is still in a high level because the reboilers of C1, C2 and (C4) still need the saturated steam to transmit the heat. Table 2 shows the reflux ratio of C4 is 11.57 and this large reflux ratio can cause the increasing of the energy requirement of the water column (C4). Besides, according to the flowsheet of this scheme, the diameter of column is so large that the emergence of many urgent cases.

In view of above these defects in the improved four-column scheme, the novel five-column heat integrated methanol distillation scheme (shown as Fig. 4) is designed in this paper to reduce the impact of these factors as far as possible by using the paralleled or combined double-effect distillation. Juntao Zhang [23] and Jinsheng Sun [25] are also proposed a five-column methanol distillation scheme by using triple-effect distillation and double-effect distillation respectively, and the effect of energy saving of the two schemes is outstanding.

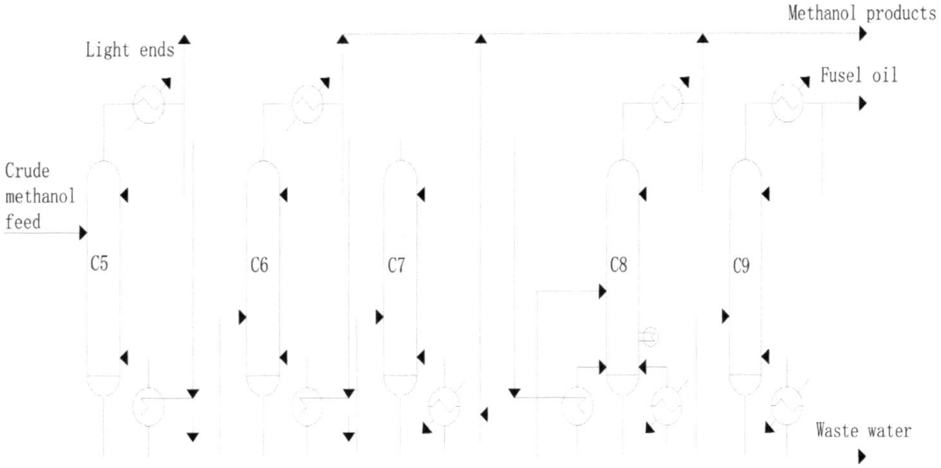

Fig. 4. The five-column heat integrated methanol distillation scheme (C5, light ends column; C6, prepositive atmospheric column; C7, pressured column; C8, postpositive atmospheric column; C9, water column)

According to the principle of the paralleled double-effect distillation, the four columns are unable to constitute the heat integration network in the improved four-column scheme. Through repeated verification, the prepositive atmospheric column (C6) is added after the light ends column (C5), meanwhile, the operating pressure of the pressured column (C7) is reduced to increase its throughput. In this way, the overhead vapor of C7 is split into three streams to respectively integrate with the light ends column (C5), the prepositive atmospheric column (C6) and the postpositive atmospheric column (C8). Due to use the paralleled double-effect distillation, the duties of the hot utility have decreased in a great margin. However, there are still part of heat from the background process is absorbed by the cooling water. So an inter-reboiler with a low temperature level is designed in the postpositive atmospheric column (C8) and it can share the heat with the background process streams, besides, we have proved the feasibility by using the column grand composite curve analysis of columns. Figure. 4 shows that the product of the top of the water column (C9) is fusel oil instead of purified methanol, which can not only reduce the reflux ratio of C9, but also make the diameter of column smaller to reduce the difficulty of operation.

Due to produce the fusel oil instead of purified methanol in the water column (C9) in the novel five-column distillation scheme, the methanol content of the feed for the water column have to limit at 0.016 wt%. In the five-column scheme, the operating pressure and the number of theoretical stage for each column are determined through the existing process, the onsite data and the requirement of the heat coupling. The proportion of the methanol products from the C6, C7 and C8 is determined by their own double-effect or heat integrated distillation system. The simulation results of the novel five-column scheme are listed in table 3.

Table 3. The operating parameters of five-column scheme

Column	C5	C6	C7	C8	C9
Top pressure (kPa)	160	110	400	110	110
Top temperature (℃)	83	66.6	104	66.6	86.5
Bottom temperature (℃)	85.6	84	118.8	104.5	103.2
Reflux ratio	1.46	1.49	2.23	1.87	4
Number of theoretical stage	33	54	54	65	29

4.2 Energy consumption and exergy loss analyses for the two methanol schemes

In the improved four-column scheme, the pressured column and the atmospheric column constitute the distillation system of heat coupling, there are three columns (C1, C2, C4) need the hot utility, and the entire duty of reboiler is 145.4×10^3 kw. As described above, the light ends column (C5) and the prepositive atmospheric column (C6) don't need the saturated steam in the five-column scheme and the total duty of reboiler is reduced to 122.7×10^3 kw. The detailed data of the reboiler duty as shown in table 4.

Table 4. The duty of reboiler in the two schemes

Column	C1	C2	C4	C7	C8	C9
Duty of reboiler (10^4 kw)	3.30	10.04	1.20	8.97	2.86	0.44

Obviously, the novel five-column scheme is much better than the improved four-column scheme in energy saving and the energy consumption of the novel scheme can be reduced by 15.61%. However, the recovery rate of methanol of the novel scheme is approximately equal with the improved four-column scheme.

Only the energy consumption analysis is not revealed the energy saving essence of the novel five-column scheme. Exergy analyses are used to further compare the two schemes, and the exergy flow rates are calculated through using the computer simulation data. The activity coefficient of the liquid is calculated using the Alcohol Package thermodynamic model. The calculation of the net work consumption of distillation column should pay more attention to the temperature of the cooling water. When the cooling water exchanges heat with other streams, the temperature of the cooling water is momentarily changed, so it is replaced the temperature of cooling water with the arithmetic average of inlet and outlet temperature of cooling water. Figure. 5 show the overall exergy loss of each column in the two methanol distillation schemes, in addition to the overall exergy loss of C6. Due to optimize the four-column scheme, the prepositive atmospheric column (C6) is added in the transformed scheme, and the overall exergy loss of C6 is 9157.6 kw. The exergy loss of each column, besides the pressured column (C7), in the novel five-column scheme is obviously less than their counterparts in the improved four-column scheme.

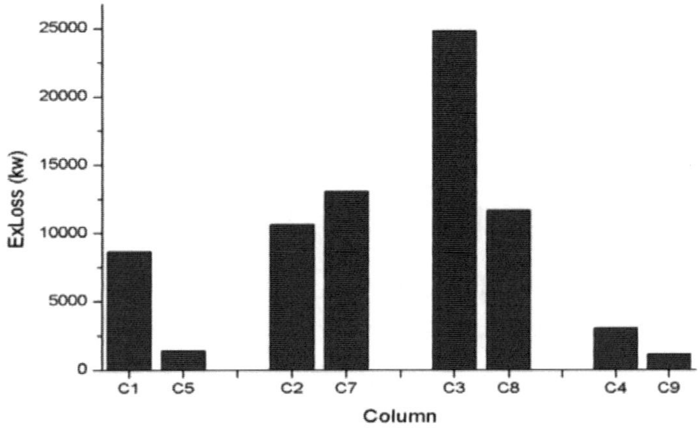

Fig. 5. The overall exergy loss analysis of each column in the two schemes

The total exergy loss of the improved four-column scheme and the transformed scheme is, respectively, 47100.1 kw and 36355.1 kw. The total exergy loss of the optimized scheme is dropped by 22.8%. Through the comparison between the exergy loss and energy consumption of the two schemes, it is indicated that the novel five-column scheme should be applied to produce the purified methanol in industry.

4.3 The novel heat coupling organosilicon distillation scheme

The primary source of the crude organosilicon monomers is the catalytic reaction products between chloromethane and silicon at high temperature and the table 5 displays the normal boiling points and mass percentage of the main components of the crude organosilicon monomers. Methyltrichlorosilane (Me1), dimethylchlorosilane (Me2), trimethylchlorosilane (Me3) and methyldichlorosilane (MeH) are the most valuable components in the crude organosilicon monomers. The purification requirements on organosilicon of these materials are shown in table 6.

Table 5. The crude organosilicon monomer feed data

Component	Normal boiling point at 1 atm/℃	Mass percentage %
LB		1
Methyldichlorosilane	41.8	4
Silicon tetrachloride	57.6	0.5
Trimethylchlorosilane	57.9	2.5
Methyltrichlorosilane	66.4	11
Dimethyldichlorosilane	70.2	80
HB		1
Temperature (℃)	40	
Pressure（kPa）	100	
Flow rate (kg/h)	26200	

Table 6. Purification demand on organosilicon

Monomer	Purification demand
MeH	99.5%
Me3	99.5%
Me1	99.5%
Me2	99.95%

The conventional nine-column scheme is simulated with the aid of existing process and onsite data to achieve the operating parameters of each column, as illustrated in table 7. The bottom temperature of the heavies removal column (C1') and the high boiling point column (C9') are obviously much bigger than that of the other seven columns in table 7. According to the principle of the double-effect distillation, the top temperature of the high pressured column is relatively high in comparison with the bottom temperature of the low pressured column, and the high pressured column can integrate with the low pressured column. Although the top temperature of C1' is less than the bottom temperature of the other seven columns, the overhead vapor of C1' still has a great opportunity to conduct heat with the other seven columns with the increasing of its top temperature.

Table 7. The simulation results of prevailing 9-column

Column	C1'	C2'	C3'	C4'	C5'	C6'	C7'	C8'	C9'
Top temperature (℃)	70.8	65.3	57.3	66.7	72.1	70.6	71.2	74.9	72.7
Bottom temperature (℃)	124.1	89.1	80.4	85.2	74.9	75.4	75.3	75.3	126.5
Reboiler duty (MW)	11.67	5.61	1.13	0.51	1.19	0.46	12.27	16.08	0.02
Condenser duty (MW)	-12.47	-5.53	-1.08	-0.53	-1.20	-0.47	-12.69	-16.12	-0.02
Top pressure (kPa)	110	180	260	220	170	150	110	110	110

According to the way of the paralleled double-effect distillation, the increasing operating pressure in the heavies removal column (C1') can make the top temperature of C1' ascending to integrate with the other seven columns. The idea is still in study to overcome some difficulties. However, we have got another feasible method, in which the operating pressure of C1' and the Me1 column (C7') is raised to 360 kPa and 280 kPa, respectively. In this way, the top temperature of C1' and C7' is increased to ensure they can integrate with the other columns, besides the high boiling point column, and the parallel double-effect or heat integrated organosilicon monomer distillation scheme is designed as shown in figure. 6 [26]. The feasibility of this heat coupling scheme has confirmed through three x-y diagrams including the heavies component and Me1, the heavies component and Me2, Me1 and Me2 [9].

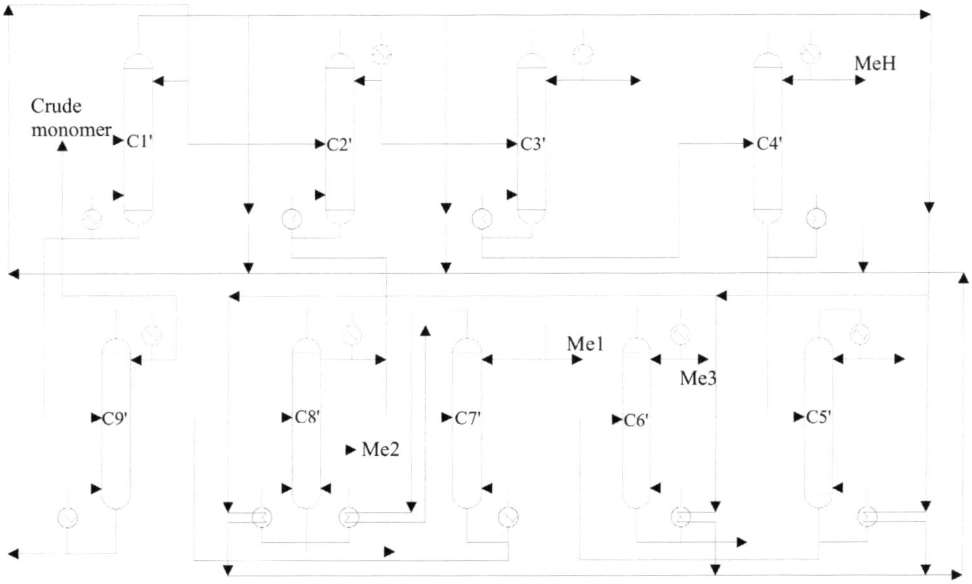

Fig. 6. The heat coupling 9-column scheme (C1', heavies removal column; C2', low removal column; C3', lights removal column; C4', MeH column; C5', azeotrope removal column; C6', Me3 column; C7', Me1 column; C8', Me2 column; C9', high boiling point column)

Figure. 6 shows the heavies removal column (C1') make the heat coupling with the low removal column (C2'), the lights removal column (C3'), the MeH column (C4'), the azeotrope removal column (C5'), the Me3 column (C6') and the Me2 column (C8'), successively. In order to gain a better energy saving effect, C8' is designed with two reboilers, the one integrated with C1' and the other integrated with the Me1 column (C7'). The operating parameters and the duty of reboiler and condenser of each column in the heat integrated 9-column are enumerated in table 8.

Table 8. The simulation results of heat integrated 9-column

Column	C1'	C2'	C3'	C4'	C5'	C6'	C7'	C8'	C9'
Top temperature (℃)	114.3	65.3	57.3	66.7	72.1	70.6	105.1	74.9	72.7
Bottom temperature (℃)	124.9	89.9	78.8	84.5	75.4	75.4	110.9	76.6	121.4
Reboiler duty (MW)	13.06	5.62	1.13	0.51	1.19	0.46	13.83	14.12	0.38
Condenser duty (MW)	-13.72	-5.54	-1.08	-0.53	-1.20	-0.47	-14.02	-14.43	-0.41
Top pressure (kPa)	360	180	260	220	170	150	280	110	110

4.4 Energy consumption and exergy loss analyses for the two organosilicon schemes

Figure. 4 indicate that each column of the prevalent nine-column scheme is not integrated with the other columns and the reboilers of all of these columns are required the saturated steam (600 kPa). According to the data from the table 7, the total duty of reboiler is 48.94 MW for the prevailing 9-column scheme. In the new nine-column scheme, C1' integrate with C2', C3', C4', C5', C6' and C8' successively and C7' integrate with C8', so C2', C3', C4', C5', C6' and C8' don't consume the saturated steam. The total duty of reboiler is 27.27 MW. Obviously, the heat integrated 9-column scheme has an absolute advantage over the prevailing 9-column scheme in energy saving, and the energy savings can reach 44.28%.

As the same with the methanol distillation scheme, the inlet and outlet temperature of cooling water are 30 ℃ and 40 ℃, and the total exergy loss is calculated through the results of computer simulation of the two schemes. The overall exergy loss comparison of each column in the two schemes is illustrated in figure. 7. In the normal scheme, the throughput of C9' is so small that the overall exergy loss of C9' is very near to zero, so it only have the exergy loss of C9' of the new scheme in figure. 7. The overall exergy loss of each column in the optimized scheme is declined in comparison with that of the corresponding column in the normal scheme except the high boiling point column (C9'). Because the crude organosilicon monomers contains some unstable components, in order to avoid these components happening thermal cracking in the heavies removal column (C1') in the novel scheme, more liquid from the bottom of C1' is pumped to C9' as the feed. That can cause the ascent of the throughput and the overall exergy loss of C9'.

The total exergy loss is 14035.5 kw and 8363.2 kw, correspondingly, in the conventional scheme and the novel scheme, and the exergy loss of the new scheme can be reduced by 40.41%. Through the above analysis and comparison, the heat integrated nine-column organosilicon distillation scheme should be a substitute for the prevalent nine-column scheme.

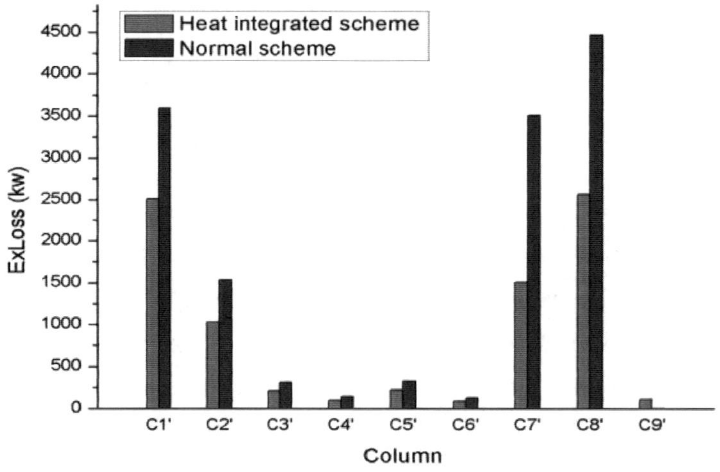

Fig. 7. The exergy loss comparison between the heat integrated scheme and the normal scheme

5. Conclusions

Paralleled double-effect or heat integrated distillation for energy saving is proposed, exampled by heat integrated five-column methanol distillation scheme and the heat coupling 9-column organosilicon monomer distillation scheme, with the aid of computer simulation. Besides, the optimized methanol distillation scheme and the transformed organosilicon scheme are comprehensively analyzed through the energy consumption and the exergy loss in comparison with the correspondingly prevalent scheme.

Adding a prepositive atmospheric column and decreasing the operating pressure of the pressured column in the novel five-column scheme can make the pressured column to simultaneously integrate with the light ends column, the prepositive atmospheric column and the postpositive atmospheric column. Besides, an inter-reboiler is added to integrate with the background process streams. The energy consumption and the exergy loss of the new scheme are 15.61% and 22.8% lower than the improved scheme.

Increasing the operating pressure of the heavies removal column and the Me1 column can make them to integrate with the other columns, in addition to the high boiling point column, to constitute a paralleled double-effect distillation system in the optimized nine-column scheme. The energy consumption and the exergy loss of the novel scheme are 44.28% and 40.41% less than the normal scheme.

The two cases expound that the key of paralleled double-effect distillation process lies in the reasonable adjustment of the energy distribution, the operating parameters and the configuration of some columns. It not only reduces the quantity of energy consumption, but also avoids the decrease of the quality of energy. For factories with a large numbers of distillation schemes of high energy requirement, the paralleled schemes lead to improve economic benefit.

References

[1] Araujo AB, Brito RP, Vasconcelos LS, "Exergetic analysis of distillation processes-a case study," Energy 2007;32:1185-93.

[2] Michele Mascia, Francesca Ferrara, Annalisa Vacca, Giuseppe Tola, Massimiliano Errico, "Design of heat integrated distillation systems for a light ends separation plant," Applied Thermal Engineering 2007;27:1205-1211.

[3] Amiya K. Jana, "Heat integrated distillation operation," Applied Energy 2010;87:1477-1494.

[4] Hilde K. Engelien, Sigurd Skogestad, "Multi-effect distillation applied to an industrial case study," Chemical Engineering and Processing 2005;44:819-826.

[5] James G. Gebbie, Michael K. Jensen, Piotr A. Domanski, "Experimental transient performance of a heat pump equipped with a distillation column," International Journal of Refrigeration 2007;30:499-505.

[6] Massimiliano Errico,Giuseppe Tola, Michele Mascia, "Energy saving in a crude distillation unit by a preflash implementation," Applied Thermal Engineering2009;29:1642-1647.

[7] Massimiliano Erricoa, Giuseppe Tola, Ben-Guang Rong, Daniele Demurtas, Ilkka Turunen, "Energy saving and capital cost evaluation in distillation column sequences with a divided wall column," chemical engineering research and design 2009;87:1649-1657.

[8] Ivar J. Halvorsen, Sigurd Skogestad, "Energy efficient distillation," Journal of Natural Gas Science and Engineering 2011;3:571-580.

[9] Jinsheng Sun, Fan Wang, Tingting Ma, Hong Gao, Yanzhen Liu, Fang Cai, "Exergy analysis of a parallel double-effect organosilicon monomer distillation scheme," Energy 2012;47:498-504.

[10] Hilde K. Engelien, Sigurd Skogestad, "Multi-effect distillation applied to an industrial case study," Chemical Engineering and Processing 2005;44:819-826.

[11] Kazuo Matsuda, Kenichi Kawazuishi, Yasuki Kansha, Chihiro Fushimi, Masaki Nagao, Hiroshi Kunikiyo, Fusao Masuda, Atsushi Tsutsumi, "Advanced energy saving in distillation process with self-heat recuperation technology," Energy 2011;36:4640-4645.

[12] M.A. Sharaf, A.S. Nafey, Lourdes García-Rodríguez, "Exergy and thermo-economic analyses of a combined solar organic cycle with multi-effect distillation (MED) desalination process," Desalination2011;272:135-147.

[13] Weiwen Wang, Wenjian Wang, "Energy Saving in Methylchlorosilane Distillation," International Conference on Computer Distributed Control and Intelligent Enviromental Monitoring 2012:290-293.

[14] Myungwan Han, Sunwon Park, "Multivariable control of double-effect distillation

configurations," J. Proc. Cont. 1996:247-253.

[15] Hilde K. Engelien, Sigurd Skogestad, "Multi-effect distillation applied to an industrial case study," Chemical Engineering and Processing2005;44:819-826.

[16] Debadrita Maiti, Amiya K. Jana, Amar Nath Samanta, "A novel heat integrated batch distillation scheme," Applied Energy2011;88:5221-5225.

[17] Anton A. Kiss, Servando J. Flores Landaeta, Carlos A. Infante Ferreira, "Towards energy efficient distillation technologies-Making the right choice," Energy2012;47:531-542.

[18] Diego C. Alarcón-Padilla, Lourdes García-Rodríguez, Julián Blanco-Gálvez, "Experimental assessment of connection of an absorption heat pump to a multi-effect distillation unit," Desalination2010;250:500-505.

[19] Diego C. Alarcón-Padilla, Lourdes García-Rodríguez, Julián Blanco-Gálvez, "Design recommendations for a multi-effect distillation plant connected to a double-effect absorption heat pump: A solar desalination case study," Desalination2010;262:11-14.

[20] Fiedler E, Grossman G. Methanol, "Ullmann's Encyclopedia of Industrial Chemistry," Weinheim : Wiley-VCH Verlag/Gmbh & Co; 2002.

[21] Meyers RA, "Handbook of synfuels Technology," New York : McGraw-Hill; 1984.

[22] Chu YZ, Qin LP, "Optimum design of methanol distillation process and columns," Chem Ind Eng Prog 2008;(27):1659-62.

[23] Juntao Zhang, Shengrong Liang, Xiao Feng, "A novel multi-effect methanol distillation process," Chem Eng Pro 2010;(49):1031-37.

[24] Jinsheng Sun, Fan Wang, Tingting Ma, Hong Gao, Peng Wu, Lili Liu, "Energy and exergy analysis of a five-column methanol distillation scheme," Energy2012;45:696-703.

[25] J. S. Sun, P. Wu, Y. H. Wang, H. Gao, L.L. Liu.CN Patent102039058 A. 2011.

[26] J. S. Sun, P. Wu, Y. H. Wang, Y. X. Su, H. Gao, J. Li, D. Yan. CN Patent 101798322 A. 2010.

[27] Michele Mascia, Francesca Ferrara, Annalisa Vacca, Giuseppe Tola, Massimiliano Errico, "Design of heat integrated distillation systems for a light ends separation plant," Applied Thermal Engineering 2007;27:1205-1211.

[28] Olga A. Flores, J. Carlos Ca´ rdenas, Salvador Herna´ ndez, Vicente Rico-Ramı´rez, "Thermodynamic Analysis of Thermally Coupled Distillation Sequences," Ind. Eng. Chem. Res. 2003;42:5940-5945.

[29] Szargut J, "Exergy Analysis of Thermal, Chemical and Metallurgical Processes," New York: Hemisphere Publishing Corporation;1988:55-152.

(45n) In Situ Transesterification of Jatropha Curcas Seed Particles with Alkaline Phase Transfer Catalyst

Subbarao Duvvuri[1], Shuhaimi Mahadzir[2] and Sintayehu Hailegiorgis[2], (1)Department of Chemical Engineering, Universiti Teknologi PETRONAS, Tronoh, Malaysia, (2)Chemical Engineering, Universiti Teknologi PETRONAS, Tronoh, Malaysia

Economic viability of biodiesel technology is at stake due to the high cost of edible oil feedstock and slow reaction rates due to limited solubility of oil in methanol. In situ transesterification of non edible oil seed particles can minimize the cost of feedstock. Microwave pretreatment of seeds can make oil molecules more reactive. Use of phase transfer catalysts can enhance the reaction rates. In the present work, these three concepts are utilized together to investigate alkaline in situ transesterification of microwave pretreated jatropha curcas seed particles in presence of benzyltrimethylammonium hydroxide (BTMAOH) as a phase transfer catalyst (PTC). It was observed that addition of PTC to alkaline catalyst has improved conversion and reduced reaction time. Microwave pretreatment of jatropha seed particles reduced reaction time drastically. The quality of biodiesel produced was within the limits of the international standard requirements of ATME D6751 and EN-14214. Thus, combined use of microwave pretreatment of seed particles, phase transfer catalysis in the presence of alkaline catalyst can increase the rate of in situ transesterification and product yield.

(45o) Catalyzed-Assisted Manufacture of Olefins (CAMOL): Updated for Use in Naphtha Service

Maximilian Walter, Petrochemicals, BASF SE, Ludwigshafen, Germany and **Suzanne Rech**, BASF Qtech, BASF Corporation, Iselin, NJ

Since the commercialization of the olefins plant, operators have searched for ways to be more competitive. One area always targeted includes keeping their furnaces in operation longer between decoking steps. Over the years, many anti-coking methods/technologies have become commercially available using different approaches to prevent coils from coking. One of the latest available technologies is the novel CAMOL™ (Catalyzed-assisted Manufacture of Olefins) Technology from BASF Qtech. As reported in the introduction of the technology to the EPC in 2008 & 2010, the CAMOL™technology differs from other available anti-coking technologies in that the technology not only eliminates the catalytic (or filamentous) coke make through a physical barrier, but also gasifies the thermal (or amorphous) coke for an even greater coking resistance. The coating process has generated a coating with almost identical thermo-mechanical properties as the base metal substrate (alloy). This allows bonding to the base metal similar to being welded with a minimum risk of spallation when going through thermal cycling. Commercially, the CAMOL Technology has been available since 2006 and is now in operation in furnaces of various designs.

With CAMOL™, operators have experienced extended run lengths, reduced energy requirement and GHG emissions, lower average TMT operating temperatures, reduced DMDS consumption, and less carburization. In addition, the technology has exhibited high operating temperature stability (>1130°C/ 2065°F).

At BASF SE in Ludwigshafen, Germany, the technology has been in operation since early 2010 in mostly naphtha service. After overcoming some initial challenges associated with a catalytic surface technology, such as reactivating the catalytic species after decoking, the coil is now successfully operating with run lengths significantly higher than uncoated coils. This paper will report on the results and learnings from the experiences gathered over the past three years along with the advancements made since the last report in 2010.

(45p) Extending Furnace Run Length and Coil Life by Optimizing Decoking

Vinod Mishra, Ingenero, Mumbai, India and **James Brigman**, Ingenero, Houston, TX

Abstract:

In the Ethylene manufacturing process, coke deposited in the radiant coils and Transfer Line Exchangers (TLEs) during normal operation is removed periodically by decoking the furnaces using a mixture of steam and air. There is a significant variation in operating parameters adopted by different plants for decoking. These are often either too conservative or aggressive, leading to production loss from increased off-stream time, TLE limitation and deterioration in coil life. This paper describes how the decoking procedure was modified based on furnace and TLE operating conditions and the decoking steps changed to obtain the best possible Ethylene yield, without adversely impacting coil life and TLE performance. This optimization has now been done in several major ethylene facilities and furnace run length has more than doubled and production has significantly increased.

SOLVING OPERATIONAL PROBLEMS IN A NAPHTHA SPLITTER

Flávio Weissheimer, GE Water & Process Technologies do Brasil Ltda, São Paulo/Brasil
Giorgio Franceschetti, Braskem, Santo André/Brasil
Rafael Henrique Camparin, GE Water & Process Technologies do Brasil Ltda, São Paulo/Brasil
Patrick Lucas, GE Water & Process Technologies, The Woodlands/US

Abstract

This paper sets out how the polymerization process of Reboilers of the (A-100) Crude Naphtha Fractionation Unit of Braskem UNIB 3 ABC can be controlled by applying polymerization inhibitors developed by GE Water & Process Technologies Ltda and Braskem´s engineering solutions.

These heat exchangers have a high degree of polymerization because of the content of olefinic compounds present in the feed composition which, when in contact with high temperatures in the heat exchange areas of reboilers, tend to polymerize. The joint solution of GE technology and Braskem engineering solutions applied to this system ensures regular operation of such equipment for periods of time greater than those hitherto recorded without manifesting loss of heat transfer efficiency.

In the 2012 shutdown while the opening of these reboilers was being inspected, it was observed that pipes had not been clogged nor was surface fouling present, which is a very positive state when compared to the historical records of previous campaigns.

The strategy adopted by GE/Braskem to ensure a longer campaign for these pieces of equipment was based on the quality control of feeding by adjusting dosages and monitoring performance using statistical analysis tools.

The use of statistical models based on Multivariable Regression Analysis (MRA) to evaluate performance, which is validated by inspecting reboilers internally, has been an efficient way to determine the degree to which the system has been fouled and also to schedule shut-downs for cleaning the mechanical parts.

Description of the Process

The (A-100) Naphta Fractionation Unit includes two distillation columns, the objective of which is to separate the light (C_6), medium (C_6 - C_8) and heavy (C_{9+}) cuts, included in the feed supply of crude naphtha.

The light cut and a portion of the heavy cut are sent to pyrolysis furnaces so as to produce ethylene. The medium cut is sent to the Hydrotreating and Catalytic Reforming Unit, with the objective of increasing the aromaticity of the system so as to produce BTX.

The A-100 processes 100 tons/h of crude naphtha, and approximately 35% of this current represents the average cut. Figure 1 below shows the schematic flow diagram of this fractionation process.

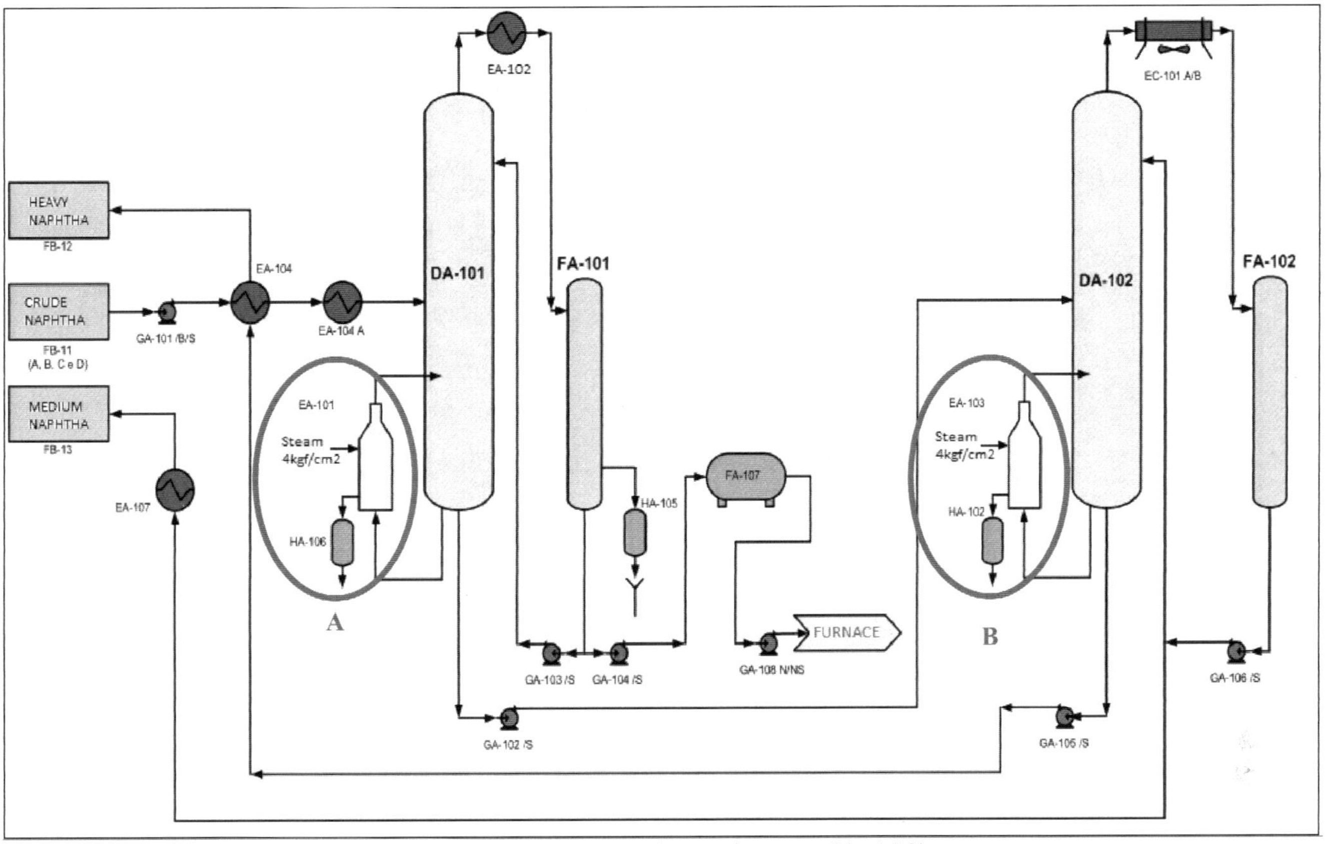

Figure 1. Process Flow Diagram (A-100)

The Reboilers, indicated by the letters A and B in Figure 1 above, operate with an injection of low steam (4kgf/cm2) to promote vaporization of the bottom stream of the A-100 distillation columns. These devices operate over a temperature range of about 150°C, which is sufficient to promote the polymerization of certain olefinic compounds present in the feed of crude naphtha.

Poor performance of these Reboilers can lead to production stoppages for maintenance and cleaning, which very often affects up to five days of production. The possibility of raising the times of campaigns and predicting the degree of fouling in downstream units helps to minimize production losses due to unscheduled stoppages.

This study will discuss the results of Reboiler A, which is the most critical equipment, with a view to synthesizing data, but gains in equipment B were also significant.

Evaluation System

Historical Record of the Campaign

Reboiler A was installed in 1972 but examination of the historical records did not reveal that the sheaf of tubes had ever been replaced. This exchanger has currently 11 tubes plugged into it.

The Campaign times of this equipment are shown in Table 1, with the last campaign being highlighted in blue so as to set it apart from the other ones. Table 1 gives a summary in the order of the timings of the campaigns, the average load processed in the period, the total load processed, the reason for operational stoppages etc.

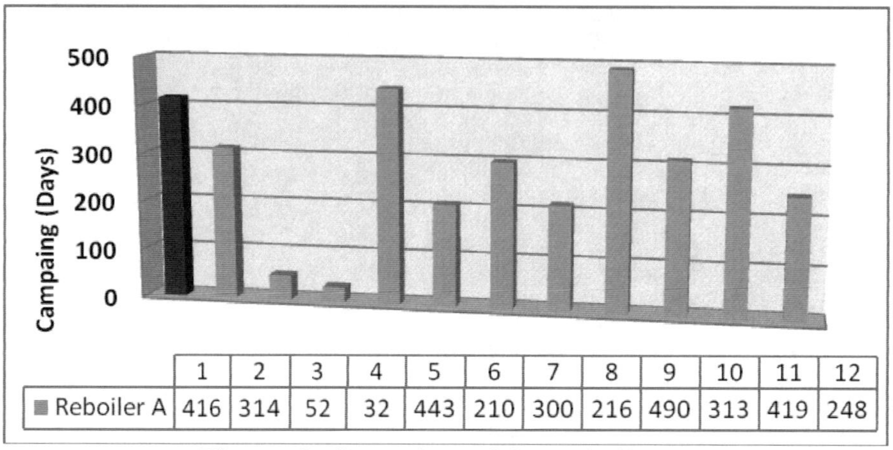

Figure 2. Campaign of the Reboiler A

Table 1. Historical Record of Reboiler A Campaigns

Date	Campaign *(days)*	Capacity *(ton/h)*	Capacity *(ton/campaign)*	Occurrence	Tubes Clogged *(%)*
13/06/12	416.0			Programmed Shutdown	0
17/03/11	314.0	93.1	701601.6	Heat exchange performance	30
03/05/10	52.0	90.8	113318.4	Stuck tube	-
11/03/10	32.0	71.1	54604.8	Heat exchange performance	50
09/02/10	443.0	72.0	765716.6	Heat exchange performance	20
16/11/08	210.0	98.1	494575.2	Heat exchange performance	-
16/04/08	300.0	104.0	748800.0	Programmed Shutdown	-
16/06/07	216.0	92.8	481075.2	Programmed Shutdown	-
10/11/06	490.0	-	-	Programmed Shutdown	-
30/06/05	313.0	-	-	Stuck tube	-
17/08/04	419.0	-	-	-	-
18/06/03	248.0	-	-	-	-
10/10/02	-	-	-	-	-

Conditions of Opening

In this review, a comparison is presented between two operational shut-downs of A-100, which occurred in March 2011 and May 2012, in order to observe the improvement in performance from the last campaign associated with the change in the GE Treatment strategy.

Campaign I: Production shutdown on 03/2011

Table 2 below summarizes the features observed during the inspection of the state of the inlet of this piece of this equipment during the shutdown of March 2011:

Table 2. Internal inspection of Reboiler A in 03/2011

Bottom Cover	Surface residue
Bottom Mirror	No deposits or internal blockage of the tubes were observed.
Top Mirror	Approximately 30% of the tubes were clogged, especially in the region of the vapor inlet, which has higher temperatures. Magnetic and non-magnetic deposits.

May/2011	**May/2011**

Figure 3. The mirror of Reboiler A before it was cleaned.	**Figure 4.** The mirror of Reboiler A at the vapor inlet before it was cleaned.

Note: It is difficult to carry out internal cleansing by Hydro jetting (pump pressure during cleaning is 20,000 psi). This reboiler has 11 tubes plugged into it, but none of them are plugged in during this shutdown.

Campaign II: Production shutdown on 03/2011

Table 3 below summarizes the features observed during the inspection of the state of the inlet of this piece of equipment during the shutdown in May 2012:

Table 3. Internal inspection of Reboiler A in 05/2012

Bottom Cover	Surface residue
Bottom Mirror	No deposits or internal blockage of the tubes were observed.
Top Mirror	Clogged tubes were not identified, not even in the region of the inlet vapor. There was only a thin layer of deposit that was easy to remove and which was evenly distributed over the upper surface of the mirror.
Spool	A thin deposit on the inner surface of the caisson.

May/2012

Figure 5. The mirror of Reboiler A before it was cleaned.

May/2012

Figure 6. The mirror of Reboiler A before it was cleaned.

Note: Hydro jetting was done quickly, without any obstructions due to deposits having been identified. This equipment has 11 tubes plugged into it, but none of them were during this shutdown.

Analysis of Deposits

The characterization of the deposits found in Reboiler A showed an 86.0% loss on ignition, thus indicating the strong presence of organic compounds in their composition, with the functional groups of carboxylic acid and ester standing out. The inorganic analysis of the sample indicated a high concentration of iron oxides and also sulfur oxides in the form of sulfides on the cover and spool of this reboiler. This is summarized in Table 4 below:

Table 4. Analysis of the Deposits of Reboiler A

Organic Compounds - LOI (%)	86.0
Water of hydration	Smaller
Sulfur compounds	Smaller
Carboxylic Acid and Ester	Greater
Dichloromethane soluble (%)	2.0
Elemental Sulfur	Greater
Dichloromethane insoluble (%)	98.0
Water of hydration	Smaller
Inorganics	Smaller
Carboxylic Acid and Ester	Greater
Composition of Dichloromethane insolubles (%)	
Carbon	56.0
Hydrogen	5.0
Nitrogen	3.0
Sulfur	8.0
Inorganic Compounds (%)	
Iron, Fe2O3 + Fe3O4	82.0
Sulfide, S2-	16.0
Phosphate, P2O5	1.0
Sodium, Na2O	1.0

Evaluation of Performance

The evaluation of performance of Reboiler A was prepared using the MRA tool, which is a GE good practice to predict fouling in critical systems. This statistical tool enables a dependent variable to be determined, in this case the percentage of opening the steam valve, by means of the numerical regression of the other operational variables of the column. This method generates a mathematical equation to predict the percentage of inlet steam when the equipment is clean, given that the equation obtained by regression includes data of this condition.

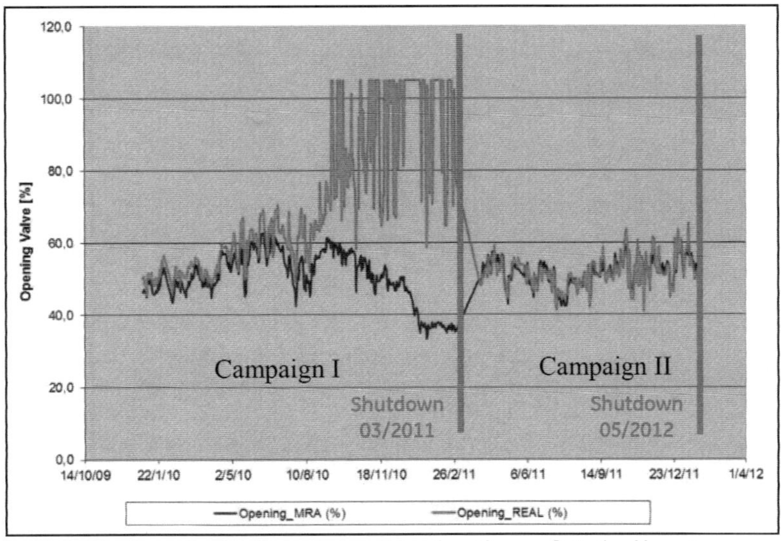

Figure 7. Performance Evaluation of Reboiler A

In Figure 7 the dark blue line represents the percentage of inlet steam when the equipment is clean, as per the equation defined by the MRA. The pink line represents the actual percentage of inlet steam. The sharp upward shift evidenced in the period between November 2010 and March 2011 shows the fouling of the system during Campaign I (prior to the 2011 operational shutdown).

After the 2011 shutdown, it can be seen that the two curves follow the same trend, without a shift, indicating that the unit has the equivalent performance of a clean piece of equipment. The change in the GE treatment strategy, taking the control of the dosages of the product Petroflo 20Y25 into account, was seen to have been effective in increasing the campaign time of this equipment.

Conclusion

The change of strategy in applying dispersant and controlling the dosages after the March 2011 shutdown, presented positive results, given the states of the inlet seen during the inspection of reboilers A and B in May 2012..

The low degree of fouling of these reboilers evidenced in the 2012 shutdown led to the production process being considered reliable, as the A-100 reboilers are no longer the pieces of equipment responsible for the loading limit of the unit. The downtime during the cleaning of the mechanical parts of the reboilers was about 10 times less than that observed in the previous shutdown, thus reducing Braskem´s maintenance costs.

In addition, the MRA statistical tool was able to produce a reliable representation of the degree of fouling of the system. This tool can be effective for scheduling shutdowns more than 03 months in advance.

References

1. Setal Lummus, "Manual de Operação da Unidade de Processamento de Nafta Bruta (Área 100)", Santo André, Brazil.
2. Ponce de Leon, Carolina; Weissheimer, Flavio and Camparin, Rafael H. (2011), "Proposta Técnica de Fornecimento de Especialidades Químicas, Equipamentos e Serviços para Tratamento Químico para a Unidade de Processamento de Nafta Bruta (Área 100)", Santo André, Brazil.
3. Kleinbaum, David G., Kupper, Lawrence L., Muller, Keith E. (2008) "Applied Regression Analysis and Other Multivariable Methods", Belmont, USA.

13AIChE
Spring Meeting, San Antonio, TX

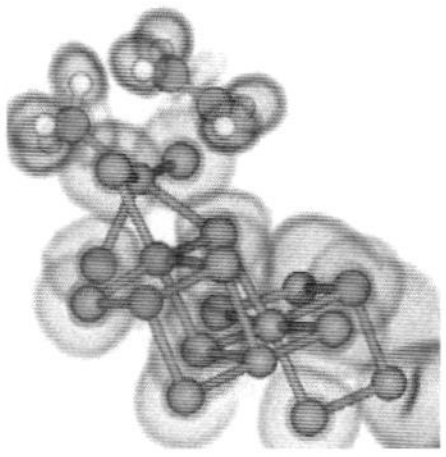

SmartProcess® Application Strategy Achieves Faster and More Efficient Control of an Ethylene Complex

Patrick Truesdale
Principal Consultant
Industry Solution Group
Emerson Process Management

April 29, 2013
San Antonio, TX

EMERSON.
Process Management

Speaker Introduction

- Principal Consultant
  Emerson Process Systems & Solutions
- BS Electrical Eng, PE
- 40+ Years experience with Refinery & Petrochem Process Control
- Joined Emerson in 2007
- Previous employers: duPont, Bonner & Moore, SETPOINT, AspenTech, Comcept Technology

Patrick Truesdale

E-mail: patrick.truesdale@emerson.com

Agenda

- Performance Package Review
- SmartProcess Ethylene Overview
  - Furnaces
  - Distillation
- Benefits

13 AIChE
Spring Meeting, San Antonio, TX

EMERSON.
Process Management
Page 3

SmartProcess Ethylene Packages

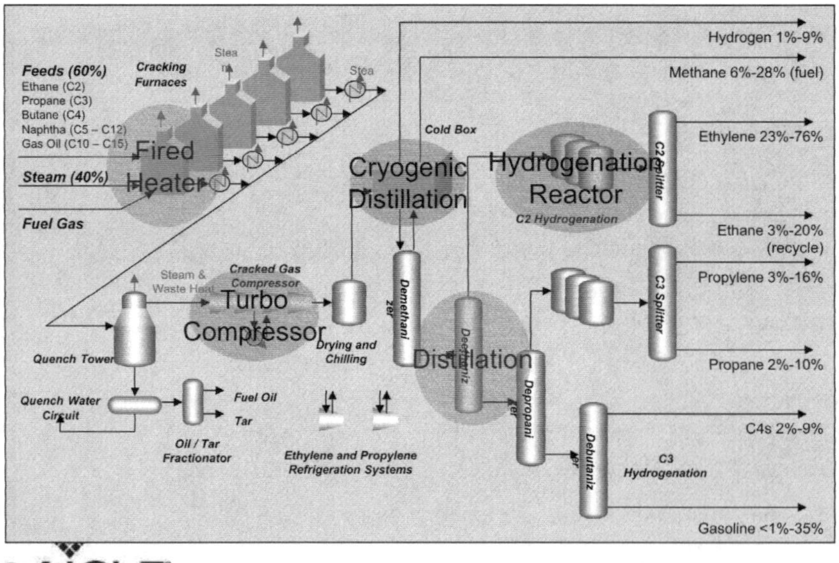

13 AIChE
Spring Meeting, San Antonio, TX

EMERSON.
Process Management
Page 4

SmartProcess Scope

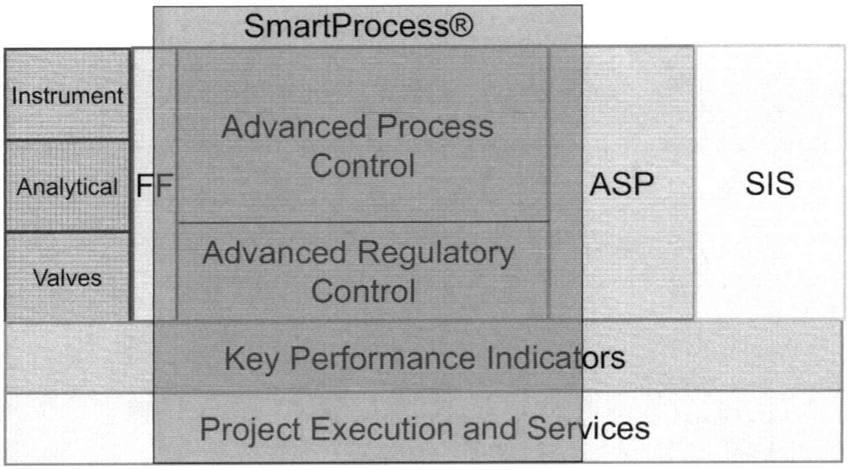

Basic Ethylene Furnace Variables

Key Operating Objectives:

- **Constant severity**
- **Maintain HC/ Stm ratios**
- **Min excess air**
- **Maintain within constraints**
- **Max charge (if needed)**

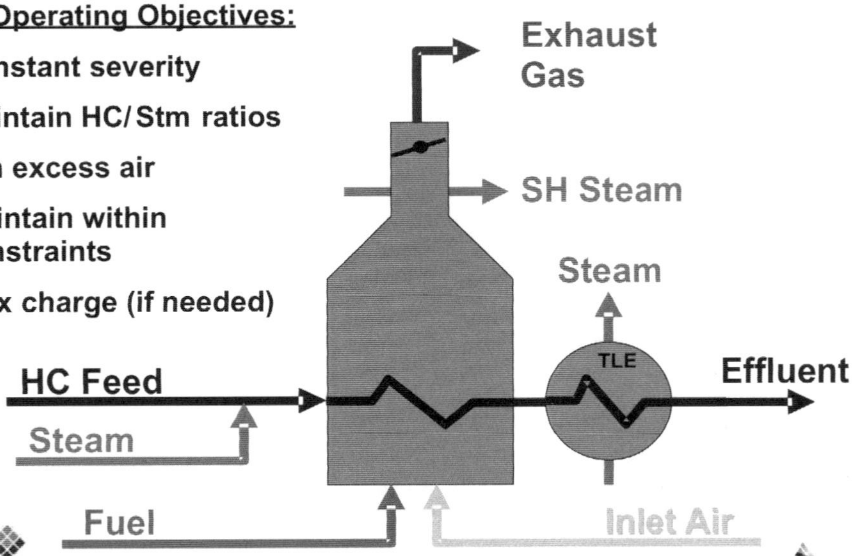

SmartProcess Furnace Module

- Design allows for different furnace configurations
  - Process Zones
  - Firing Zones
  - Multiple TLE analyzers
- Advanced Regulatory
  - Charge Control
  - Steam Control
  - Severity Calculation
  - Combustion Control
  - KPI's
- Advanced
  - Pass Balancing
  - Heat Balancing
  - Severity Control
  - O2 Control

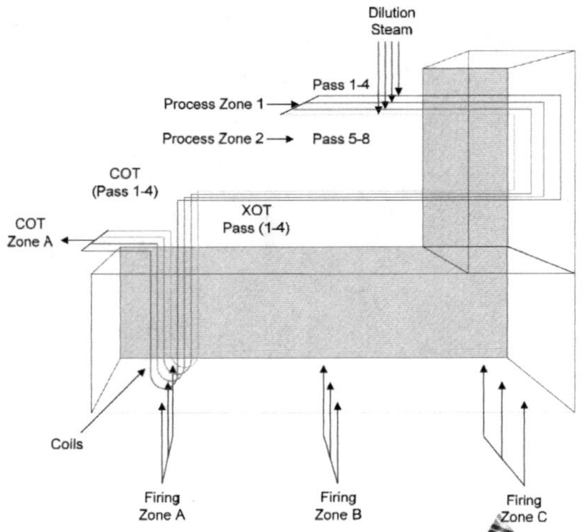

Standard Furnace Module Design

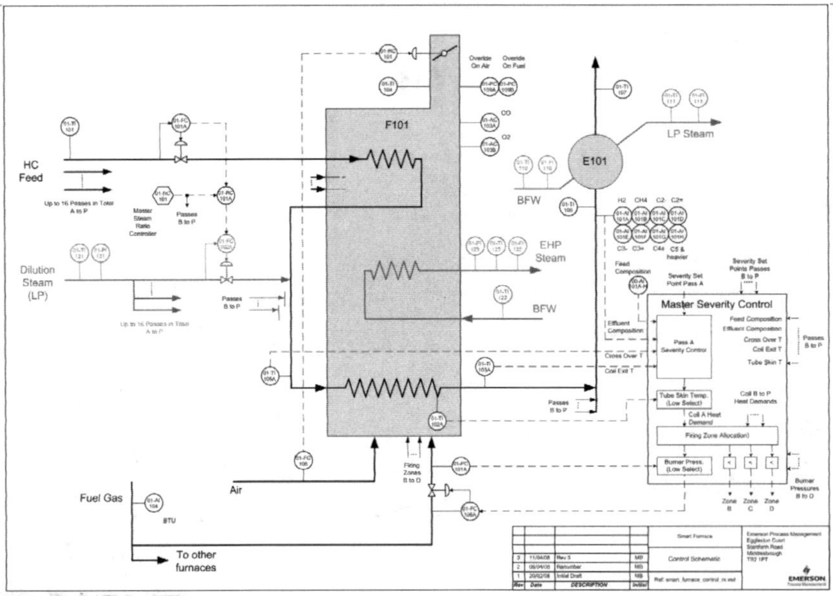

Charge Rate Control

Functions:
- Pass Balancing
- Constraints
- Feed Maximization

Functions:
- Ramping
- Limits & Alarms
- Multiple Feed Types
- Furnace Status
- Pass Offsets

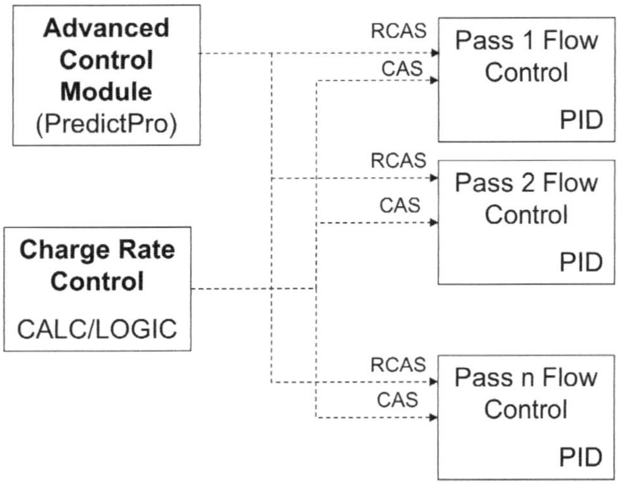

13 AIChE
Spring Meeting, San Antonio, TX

EMERSON
Process Management

Steam Ratio Control

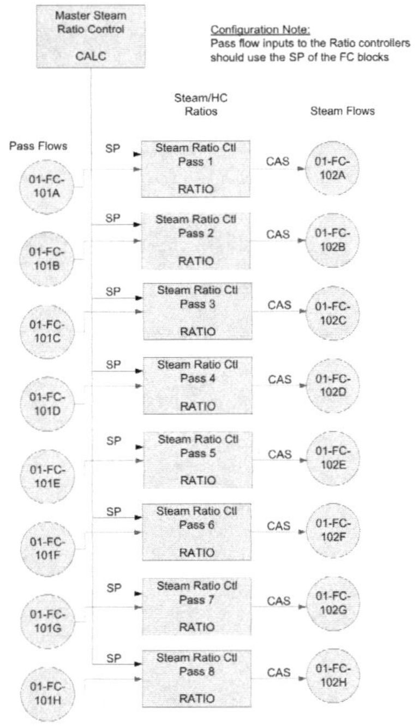

13 AIChE
Spring Meeting, San Antonio, TX

EMERSON
Process Management

Severity Control

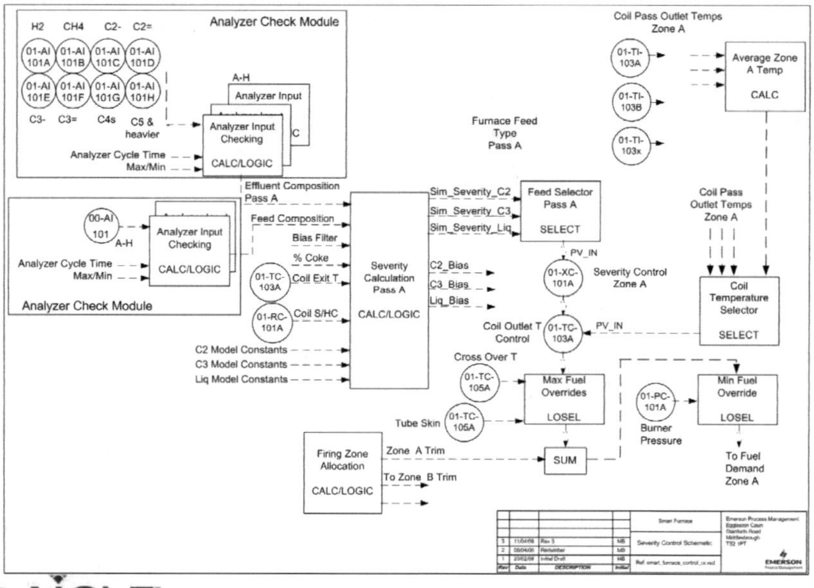

13 AIChE
Spring Meeting, San Antonio, TX

EMERSON. Process Management

Automated Furnace Sequences

- Standard module templates to be used for Furnace Startup, Shutdown and Decoking sequences.
- Utilizes preconfigured Sequence Function Charts (SFC) running in DeltaV controllers or App Station
- Standard sequence functions can be customized to match site requirements
- Manual operator steps with confirmation can be mixed with automated steps
- Sequence execution control graphics and faceplate templates

13 AIChE
Spring Meeting, San Antonio, TX

EMERSON. Process Management

Key Performance Calculations

- Furnace efficiency
  - Heat loss method
- Furnace Duty
- TLE Duty, Fouling factor
- Stack gas estimated composition
- Stack gas daily total tons exhaust for CO, CO2, S, N, etc.
- Stack gas NOx emissions (optional – based on CEMS)

Distillation - Classic Multivariable Problem

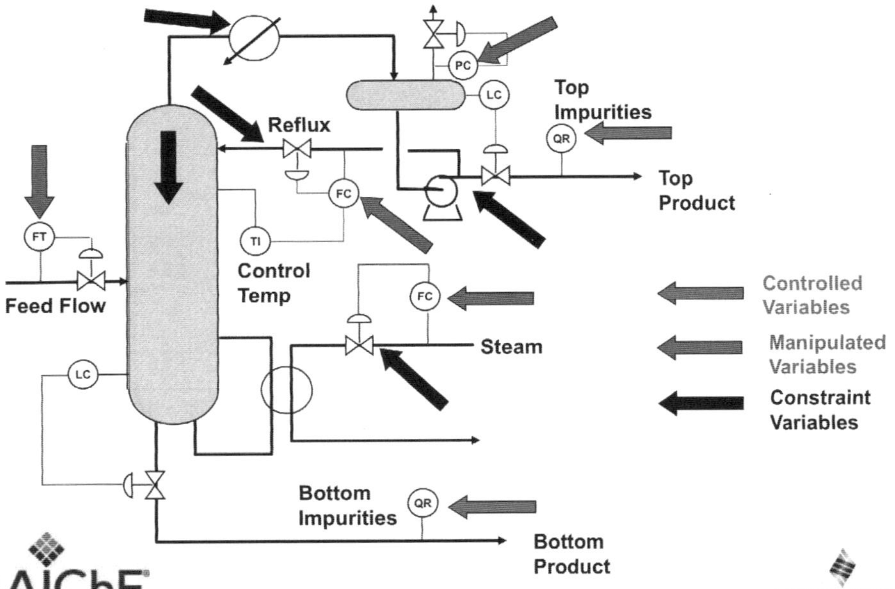

Distillation Control Module

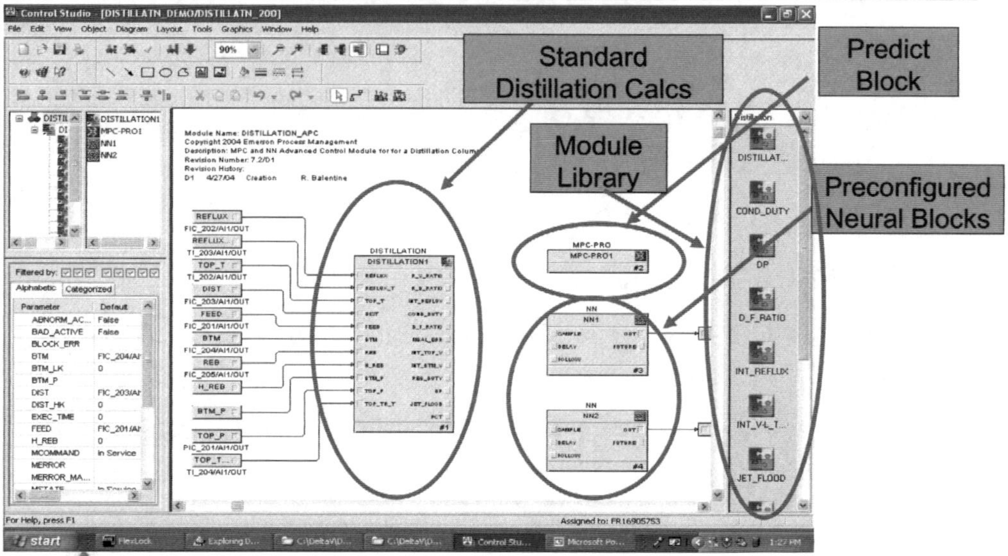

13 AIChE
Spring Meeting, San Antonio, TX

EMERSON
Process Management

SmartProcess Distillation

- DeltaV PredictPro with up to 4 MV's
- DeltaV Neural for up to 2 product qualities
- Standard set of calculations & constraints
  - Internal reflux
  - Pressure compensated temperatures
  - Jet flooding
  - Reboiler & condenser duty
- Engineering Services to design, implement and commission

13 AIChE
Spring Meeting, San Antonio, TX

EMERSON
Process Management

SmartProcess Applications

- Licensed industry applications for common processes that are:
  - Pre-engineered, tested, documented
  - Designed by APC Consultants with extensive practical experience
  - Supported & enhanced
- Bundled combination of:
  - DeltaV libraries: composite blocks, modules & templates
  - DeltaV simulation and demonstration example
  - APC tools (e.g. Predict, Neural)
  - Application code (e.g. VB, .Net)
  - Other databases and apps (e.g. SQL*Server, Excel, TankMaster)
  - Sample DeltaV Graphics
  - Documentation
- Templates to be used to configure for specific site requirements
- May be implemented by Emerson or Customer

13 AIChE Spring Meeting, San Antonio, TX

EMERSON. Process Management

Benefits

13 AIChE Spring Meeting, San Antonio, TX

EMERSON. Process Management

Ethylene Furnace High Temperature Imaging Cameras to Monitor Burner Flames ,Tube Temperature and Tube Failure Detection

Thomas M. Canty, P.E.
President
J.M. Canty Inc
Buffalo, NY USA

Abstract

The development of explosion proof, high temperature imaging cameras which operate in ethylene furnaces without cooling air now allows monitoring of ethylene furnaces to improve energy/operating efficiency along with providing important safety benefits. Imaging cameras work off the visual wavelength which provides 2 dimensional view and temperature measurement of the tube temperature which is more accurate than thermocouples and IR probes. In addition visual images of the furnace burners lets you detect flame detachment, smoking burners and flame shape in order to increase burner efficiency and save energy. The safety aspects of the Camera imaging systems is significant as it allows visual detection of pin hole leaks and tube failure in the furnaces.

Introduction

Ethylene furnaces, or any type of cracking unit, rely on thermal management in order to make product efficiently and to operate in a safe manner. Having 'eyes' in the furnace can help to identify problems in their early stages of development. High temperature camera technology is able to survive the environment and give a wide angle view of the tubes from which temperature and tube blockage may be determined. Achievable benefits from a camera system include:

1. Temperature analysis capacity for all areas of the image, not just single point. Multiple zone capability for user customization of analysis.
2. Flame detection and size measurement
3. Detection of tube movement and ability for software to automatically move measurement zones as tubes distorts.

Imaging Fundamentals

Basics of an imaging system include a high resolution sensor, proper illumination, appropriate hardware for the process and a robust software analysis package. For this particular application illumination is created by the process and need only be managed optimally for best results. A high density pixel array yields resolution of the image and measurement as well. The hardware and software are the more challenging aspects of the application and will be discussed in some detail.

Hardware

The design of the camera system is unique in that it's internally exposed lens assemblies are **designed to survive without cooling air**. When installed properly, the combination of system materials and heat transfer coefficients at the camera tip limit heat transfer to the system and allow it to run hot without chance of damage. This feature, when system is installed in a clean, gas burning environment, does not disrupt the temperature or air balances of the furnace because no cooling air is being introduced to the furnace. In addition, the design includes a disposable, quarts shield which protects the process exposed lenses from direct contact with the internal atmosphere. Should any particulate contact the system, the quartz shield is there to protect the optical lenses. Replacement of the protective quartz shield is inexpensive. Replacement of damaged lenses is very expensive and so the Canty design guards against that possibility.

Internally, the system is a two piece assembly consisting of the internal insertion lens and the external camera package. As mentioned earlier in this document, Canty employs a fused glass window as part of the electronic housing package. This has a dual purpose; Firstly, the camera housing assembly is a separate entity which carries WP and EXP ratings. Secondly, the camera housing bolts to the internal lens system and transmits the optical rays through the fused glass window to the CCD. This mounting also creates a hermetic barrier between the process and the electronic housing thereby preventing migration of hot, process gasses outside the pressure boundary of the system.

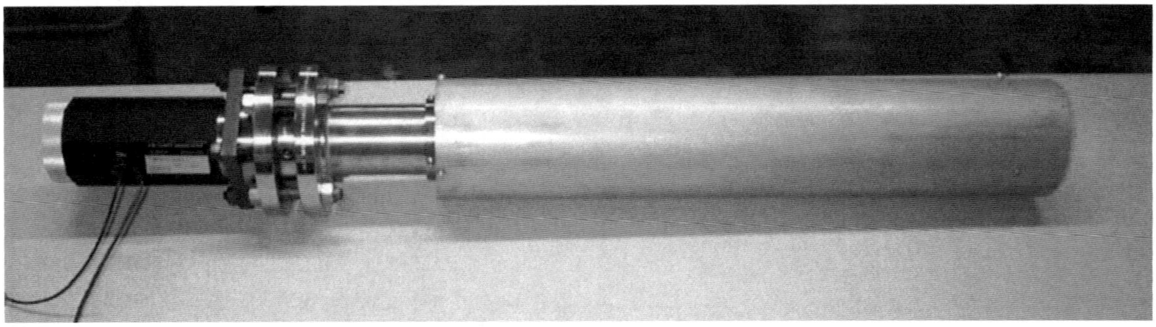

Figure 1. High Temp Camera Assembly

The camera electronics come with a full array of controls that allow it to be optimized to the software function to achieve accurate, consistent data output.

As previously mentioned, the installation is an important feature of the design. When installed correctly, the camera sits inside the refractory (see Figure 2) where its exposure to the process environment is minimized thus minimizing heat transfer into the internal lens.

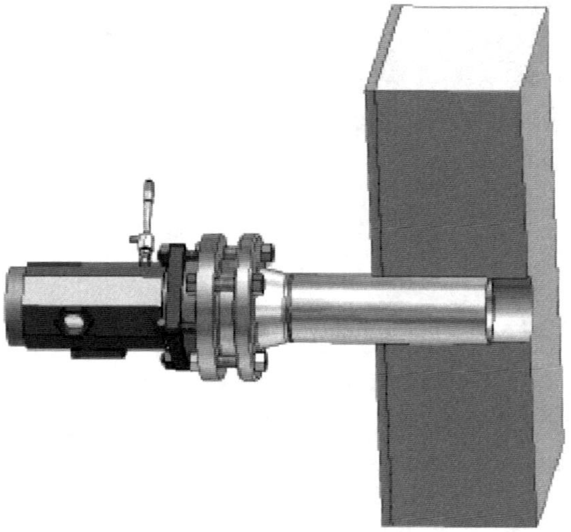

Recessed into Refractory

Figure 2. Camera Installation Schematic

Software

With respect to software design, the Canty software is fully flexible in analyzing any portion of the captured image. Multiple analysis zones can be drawn to monitor discreet areas of the view and each zone can run its own particular algorithm depending on need. The benefit here is that multiple tubes can be continually surveyed for changes and trends. The analysis data is also not single point, but is comprised of all the data points in the analysis zone. By choosing proper software filters various information can be obtained from the full array of data such as hot or cold spots. This data can also be archived in total or as snap shots at determined increments of time. The on screen data output can also be updated in the short or near term and can thus be used to judge what is happening in the furnace over time with just a glance at the screen. For instance, using the histogram feature with a 1x daily output a long term trend can be presented on screen in real time.

The basic theory of temperature measurement is based on the radiative emissions of a body at temperature – as the body temperature increases, the radiation curve shifts towards the shorter wavelengths and that is why as a body heats up the eye cannot detect it until it reaches 1000F which is the approximate visible point where emitted radiation is dark red in color. As the temperature increases, the color graduates to bright red, then orange then toward white. A pyrometer functions by measuring a narrow band of wavelength intensity and comparing it to the intensity of a black body to determine the temperature. From the graph below it can be seen that if a single wavelength is monitored the intensity of that measurement changes as the body temperature changes (curve shifts) which then corresponds to a different temperature on the calibration scale.

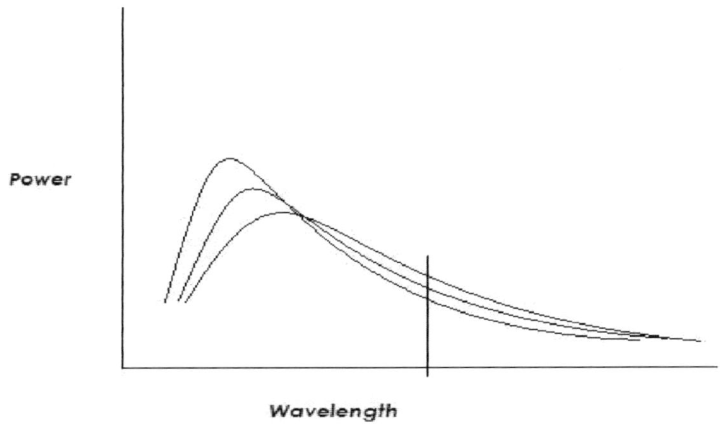

Figure 3. Power Curve at Varying Temperatures

The same is true of a camera system measuring the increasing intensity of an object as it radiates from approx 1000 F (dim, dark red color) on up the temperature scale to 3000+ F for instance which would be a bright white. From the graph above one can see that as color goes from red toward white the peak intensity is always increasing and it is the measure of this intensity which indicates the object temperature. Using an RGB camera the measurement can also be made by monitoring the individual color components of the signal impinging on the chip. For industrial uses, the measurement of intensity usually suffices.

Standard instrumentation used in industry include thermocouple type devices (contact) and thermal imaging and pyrometer systems (non contact). In severe temperatures and atmospheres, the contact devices deteriorate quickly and provide gradually less reliable information leaving the operator running blind. They also create a discontinuity on the surface they are welded to which, in the case of a furnace tube, can lead to premature tube failure.

Non contact systems have difficulties with calibration drift and emissivity adjustments to account for the fact that many items being measured are not black body emitters. In addition, these technologies are single point measurement systems and are best used as hand held devices which can be difficult to use and provide subjective measurement results based on operator preferences for targeting.

Vision technology offers several benefits these other technologies do not. The visual aspect of the system allows the operator to view the process and confirm the readings received. It also allows flexibility to set up measurement zones on screen at any time to assess any or all areas of the process in view should a change be required. Since vision systems are generally fixed installations, an added benefit is that the information is taken from the same process points and features within a given measurement zone at all times which means there is no error in sighting of the instrument.

The vision system is also unique in that it does not suffer calibration drift as almost all instruments do. Once the calibration is set the readings repeat themselves within a couple degrees F. The calibration is based on thermocouple or pyrometer readings initially, however after that calibration is finished it does not have to be redone in the future as long as the system remains in its location.

Process history and performance can also be best documented by a vision system. First off, the camera can analyze any area of process that is within the systems view. This affords the operator the capability of viewing many points of interest instead of just single point temperature readings. As an example, if the system is monitoring a tube bank it can assess the temperature of the tubes for the full length of view. If there is a local flow problem in a tube it will become overheated in one section. The camera can locate that section and indicate its temperature rise over the rest of the tube or adjacent tubes. This information allows the tube location to be identified as well as the position in the tube that has encountered the upset condition. Over time this historical data, recorded in video or still image format by the control software, may be significant in improving the process. In addition to the ability to monitor the full area of the view, the full area of view can also be segmented into multiple zones of measurement, i.e. each tube having its own measurement zone for automated process control capabilities. This capability allows the tube to be operated at optimum efficiency since the addition of excess heat can be avoided thereby reducing energy draw to operate the furnace.

Thermal Analysis

Tube Temperature

The following figures illustrate the temperature analysis capability for tubes in a live test situation. Measurement zones are placed over the tubes (figure 4) as the furnace is heated and cooled. The temperature plots are shown in figures 5 and 6. The results are as expected and uniform from tube to tube, although a small number are included in this demonstration.

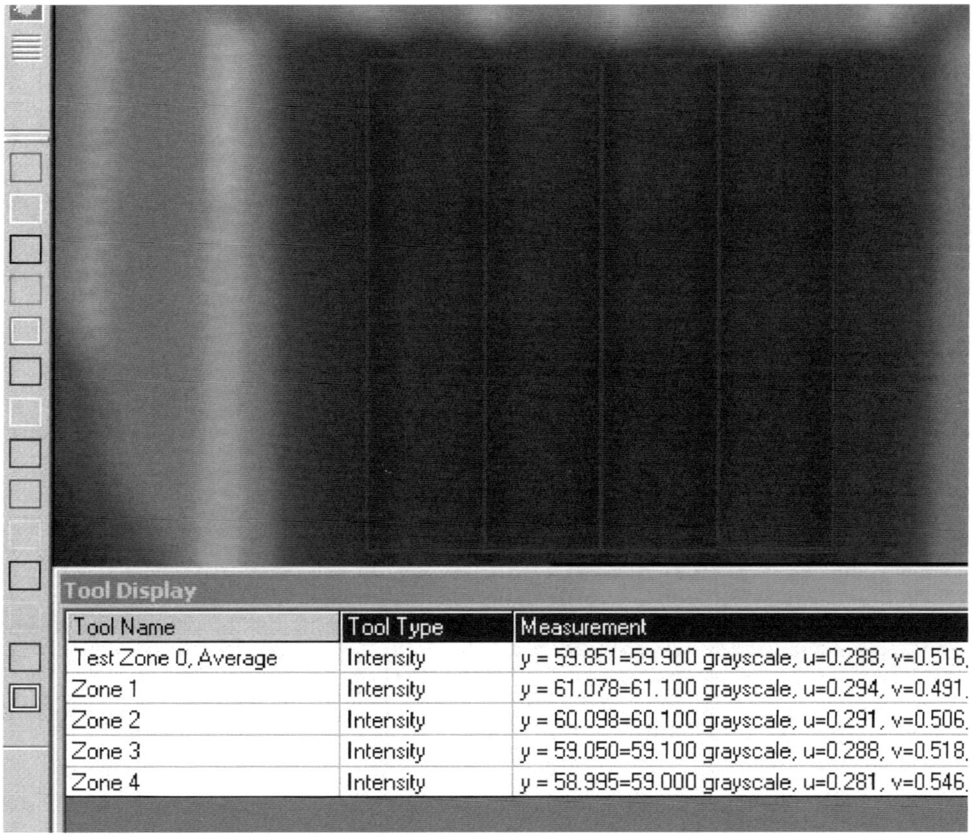

Figure 4. Software Screen Shot

As thermal loads are fluctuated up and down the following graphs track those changes. The cooling cycles are left to natural convection and thus are not linear in nature.

For the following curves, the vertical scale is gray scale intensity which is measured 0 – 255 (pitch black to bright white). The hotter the tube, the higher, or more intense, the gray scale reading. This value in process would be calibrated to the temperature. The horizontal scale is the image number, or time.

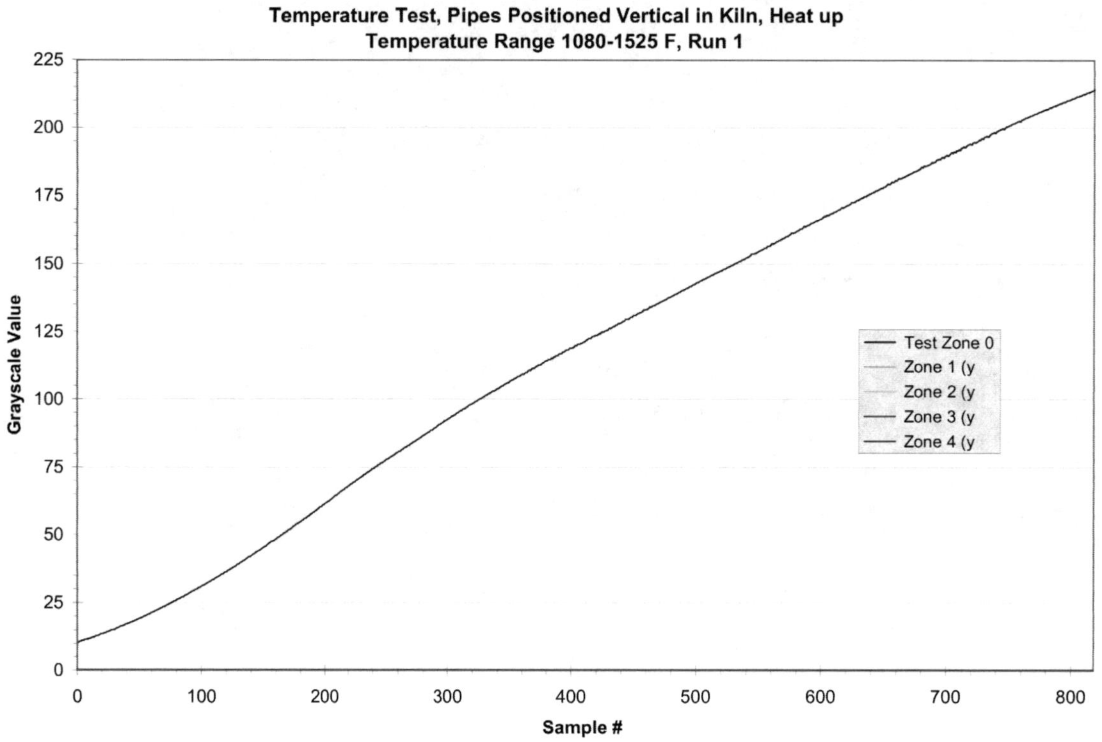

Figure 5. Heating Cycle

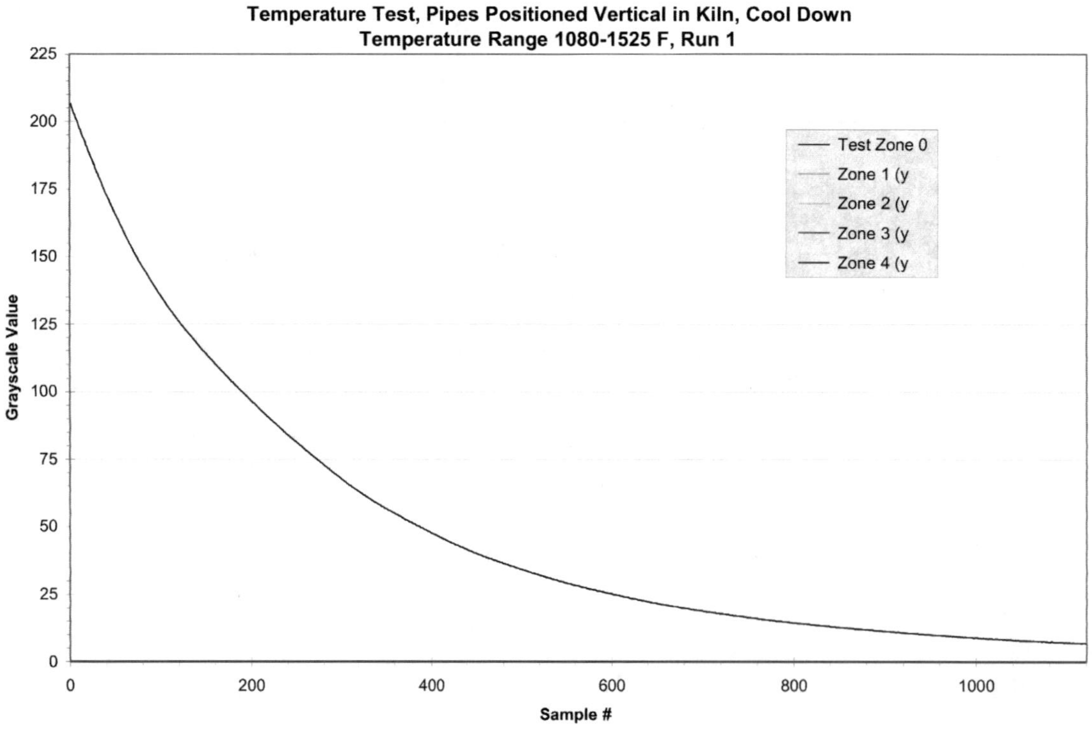

Figure 6. Cooling Cycle

Flame Measurement

Flame temperature can be monitored in the same way as described previously, however the flame size, or presence, can also be monitored. Figure 7 shows a live flame and Figure 8 shows a digital capture. Computing the flame area, or shape, can be done to track its changes over time.

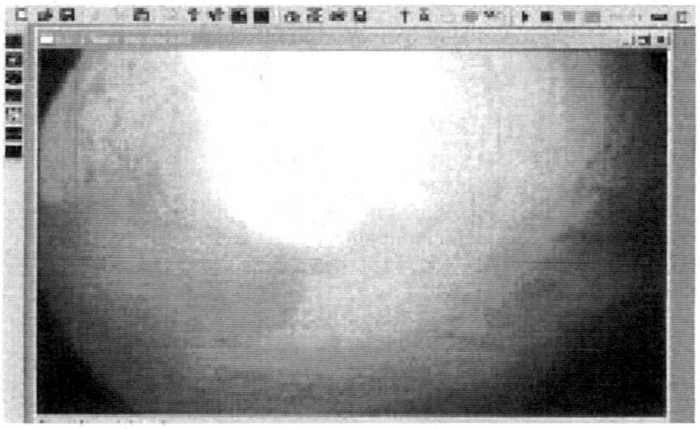

Figure 7. Kiln Flame (Live)

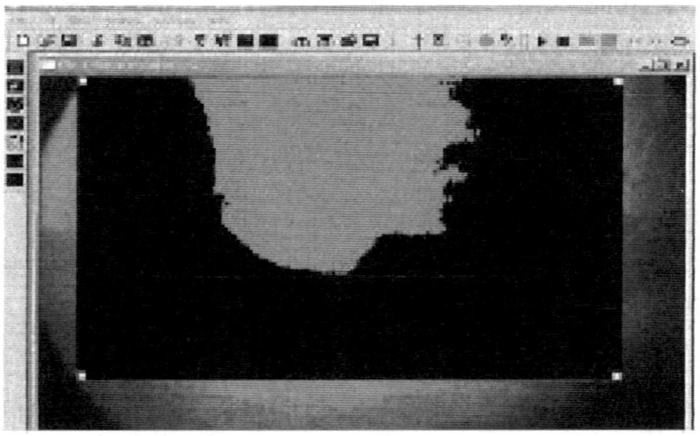

Figure 8. Kiln Flame (Digital)

Conclusion

Vision Technology presents an opportunity to increase effectiveness of furnace monitoring through hardware and software capabilities. The thermal design at the core of the hardware enables the system to provide the operator a view into the furnace and feed back on its thermal performance over time. This data can assist in detecting building upset conditions, but also provide a record for study to assist in improving the operational method of the furnace.

(45v) Straighten up! Non-Linear Compensation Using Feed Forward Control

Barbara Hamilton, Process Systems and Solutions, Emerson Process Management, Hampden , ME

Non Linear processes occur throughout the petrochemical plant. While the PID loop is surprisingly robust for most situations, there are cases where compensation of these non-linearity's is critical for stable control. Similarly, disturbances can cause major disruptions in control stability.

A method of feed forward control, using data based models, can provide robust control throughout the process range. Historical data are used to build models to drive the output toward the solution with PID control providing the final drive to zero error.

The model is also used to develop dynamic tuning. The result is a simple to configure, robust solution to manage non-linear loop and/or disturbance variations.

(45w) Adsorption of CO2 by Alkaline Activated Ca-Al2O3-SiO2 Material

Naim Faqir, King Fahd University of Petroleum & Minerals, Dhahran, Saudi Arabia and Reyad Shawabkeh, Chemical Engineering, King Fahd University of Petroleum & Minerals, Dhahran, Saudi Arabia

Natural gas contains a slight level of carbon dioxide which leads to several industrial and domestic heating problems. In addition, it is one of main greenhouse gases that contribute for global warming, where its pollution is a wide-spread problem arising from a wide range of industries, leading to complex contaminant in atmosphere. Given the increasing pressures worldwide to minimize the emission of greenhouse gases into atmosphere, CO_2 must be remediated efficiently and cost effectively.

In this short abstract a chemically treated limestone and alumino-silicate material were utilized for removal of CO_2 from natural gas using a combined adsorption-chemical reaction processes. This functionalized Ca-Al_2O_3-SiO_2 material was characterized according to its crystalline phase, surface area, pore size distribution and surface morphology. Then it was tested against different concentrations of CO_2 in CO_2-N_2 gas streams.

Pressure Gauges: Hidden Danger or Visible Safety

by Jeff Placek, WIKA Instrument, LP, Lawrenceville, GA

Introduction

In recent years, refining plants, petrochemical plants and chemical plants have been required to increase capacity, but have been provided with very few upgrades to existing equipment to accommodate the increased capacity requirements. This has caused plants to have increased stress and strain on their systems and operating units. For example, higher pulsating pressures and heavy mechanically-induced vibrations are translating to all areas of the processing equipment, especially pressure instrumentation.

Pressure gauges are one of the most important, yet most overlooked or ignored keys to ensuring the continued safety, productivity and profitability of process plants. Process pressure gauges are used to read pressure on transfer lines, discharge pumps, and other places where it is important to be able to read pressures.

Problems Associated with Misapplied, Faulty or Ignored Gauges

Incorrect, misapplied, damaged, faulty or ignored gauges can lead to a number of issues, including safety problems, downtime, and costly regulatory agency fines and other penalties.

Process Safety:

One group of problems relates to process safety issues, such as loss of containment that leads to fugitive emissions, fires and explosions. Bourdon tubes in malfunctioning gauges can rupture, causing the release of media, such as gasoline, oil or gas. Because this media is often volatile and corrosive, it can cause fires, explosions, or other serious safety issues, such as a toxic gas releases or acids spraying onto workers.

Media releases are common results of misapplied, faulty or ignored gauges. Fires and explosions are less common but they do occur. When reviewing plant instrumentation, findings have revealed instances where the wrong gauges have been installed in locations on pumps. In one instance, a gauge vibrated so severely when the pump came out of alignment that it caused the Bourdon tube in the gauge to crack. The media coming out of the tube leaked onto the pump, which was excessively hot because of age and misalignment, causing a fire.

Expensive Downtime:

A second group of problems relates to expensive downtime, which can result from the breakdown of process safety just mentioned, or simply from having to shut down a process in order to replace incorrect, misapplied, broken or otherwise damaged gauges.

1 - Safety-Related Downtime:

As a result of the problems identified above (emissions, fires, explosions, etc.), there can be any number of causes of subsequent downtime, including fire suppression activities, released media containment, line shutdown, personnel evacuation, and then eventually having to rebuild or repair the lines, replace equipment, etc. In addition, if there is a media release with no injuries or fatalities, the plant is required to notify the EPA. However, if there is a media release that does cause injuries or deaths, the plant is required to notify OSHA. These can lead to very long process shutdowns, anywhere from a day to a month or more. (More on OSHA and the EPA below.)

2 - Non-Safety-Related Downtime:

When a line fails, causing a gauge to fail, even if there is no release or safety incident, downtime can still occur as a result of the process needing to be shut down in order to replace the gauge. Gauge failure rarely causes other parts of the process to fail, although gauge failure is usually an alert that there is a problem. In most cases, a plant worker will simply replace the pressure gauge and move on. When another failure occurs, the worker will do the same, until the root cause of the failure, such as a misaligned pump, causes the plant to shut down the line, resulting in more costly, unplanned downtime and a major maintenance time.

Regulatory Agency Fines and Penalties:

A third group of problems relates to the potential for significant fines and other penalties from federal regulatory agencies (primarily OSHA and the EPA), as well as from state and local jurisdictions, that result from compromised process safety.

1 - OSHA:

OSHA addresses many process-related safety issues in 1910 Subpart H, "Hazardous Materials." More specific is 1910.119, "Process safety management of highly hazardous chemicals." In its Process Safety Management section, OSHA notes that, "Unexpected releases of toxic, reactive, or flammable liquids and gases in processes involving highly hazardous chemicals have been reported for many years in various industries that use chemicals with such properties. Regardless of the industry that uses these hazardous chemicals, there is a potential for an accidental release at times they are not properly controlled, creating the possibility of disaster."

OSHA goes on to note that: "To help ensure safe and healthful workforces, OSHA has issued the Process Safety Management of Highly Hazardous Chemicals standard (29 CFR 1910.119), which contains requirements for the management of hazards associated with processes using highly hazardous chemicals."

OSHA's standard emphasizes the management of hazards associated with highly hazardous chemicals and establishes a comprehensive management program that integrates technologies, procedures, and management practices.

While OSHA does not monitor gauges in specifically, they will enforce guidelines for worker safety. OSHA investigates injuries and deaths, and may levy fines or other penalties.

2 - EPA:

According to Greg Neneman, director of business development solutions for Valin Corporation (San Jose, Calif.), fugitive emission regulations from the Federal level have tightened up over the past several years. In many states, such as California, Oregon and Washington, the regulations are even tighter.

"One challenge companies face is ensuring compliance with emission regulations when they upgrade or expand their processes," points out Neneman. "They are not allowed to produce additional emissions beyond the original limit set by regulations."

From the standpoint of federal EPA regulations, each company that produces or emits volatile organic compounds (VOCs) is required to come up with a leak detection and repair program. "In some areas, county and local requirements are even tighter than federal requirements," continues Neneman.

In certain areas, companies that produce or emit VOCs are required by regulation to do audits on valves, flanges, gauges, and all other process connections on a quarterly basis. "And in some jurisdictions, the regulatory agencies can give plants 24 hours notice before they descend on that plant to check all of their records and conduct other investigations," adds Neneman.

In May 2012, the U.S. Environmental Protection Agency (EPA) levied a $1 million civil penalty on a plastics company for alleged violations of the Clean Air Act at two of its chemical manufacturing plants. In addition to paying the penalty, the company agreed to implement a comprehensive program to reduce fugitive emissions of hazardous air pollutants (HAPs) from leaking equipment such as valves and pumps. The settlement also required the company to implement enhanced work practices, including more frequent leak monitoring, better repair practices, and new efforts designed to prevent leaks. After being fined by the EPA, the company realized that instrumentation was one of the areas where they having fugitive emissions. Until that time, workers had been replacing gauges left and right, because they weren't aware of the problems that could occur. The gauges weren't a big expense to them, so they considered the replacement activity to be more of a nuisance than anything. It's when they had the rupture that they realized there was a real problem.

An audit of the plant, which involved reviewing at all of the instrumentation, found that in many applications, they were using the wrong instrumentation. To correct the issues, the right instrumentation was specified for them, and the correct gauges were ordered. The plant now has its instrumentation under control.

While conforming to EPA regulations has always been a challenge for process facilities, it is going to become even more challenging in the future as a result of EPA's recent introduction of new requirements related to fugitive emissions and leak detection and repair (LDAR) strategies. On April 17, 2012, the EPA Administrator signed a suite of final air quality regulations that affect both upstream and midstream oil and gas sectors - the production/processing and transmission/storage sectors of the oil and natural gas industry.

For the first time, EPA will regulate air emissions from natural gas wells that are hydraulically fractured, as well as other emission sources associated with exploration, production, processing, and transportation of oil and natural gas.

Promulgation of the final rules will result in new control equipment requirements, as well as increased requirements for monitoring, recordkeeping, and reporting for owners, operators, contractors and vendors.

The final rules require the use of emission controls, as well as work practices, through two different air regulatory programs. One is the New Source Performance Standards (NSPS). The other is the National Emissions Standards for Hazardous Air Pollutants (NESHAP).

EPA finalized a new NSPS at 40 CFR Part 60, Subpart OOOO, which regulates emissions of volatile organic compounds (VOC) and sulfur dioxide (SO_2) from a variety of oil and gas exploration, production, processing, and transportation facilities, including natural gas wells, certain compressors and pneumatic controllers, storage vessels, and sweetening units at onshore natural gas processing plants. Subpart OOOO also includes startup, shutdown and malfunction provisions.

The EPA has also made it clear that oil and gas production and processing sources will be the focus of its National Enforcement Initiative (NEI) for fiscal years 2012 and 2013. With the new NSPS and revised NESHAPs in place, EPA now has the tools it needs to enforce in a comprehensive manner.

As a result of these new regulations, plants should look at everything more closely, including gauges.

How Problems Can Occur With Gauges

Technical problems can occur with process pressure gauges in a number of ways, which can be conveniently categorized into seven broad categories with specialties or subsets. The most common reason for gauge problems is that workers often install gauges without knowing how they should be installed or accessorized. As a result, the gauges can fail because they were never designed to function in those specific situations. They can also fail because they were installed incorrectly, or because they were not properly maintained.

One way to eliminate or minimize these problems is to consider specific pressure gauge accessories. These can prove to be the extra step needed to ensure a successful, proactive preventive maintenance program, and for extending the life of the gauges. Accessories can provide additional protection for the gauges during their lifetimes.

1 - Mechanical Vibration.

This is oscillating, reciprocating or other periodic motion of a rigid or elastic body or medium forced from a position or state of equilibrium. Typically, vibration is the main cause of pressure gauge failure in process plants. Pump applications produce most of this vibration.

Gauges that aren't designed to withstand vibration may be installed in applications where vibration occurs, causing the gauges to come out of accuracy. Even more serious, a gauge can vibrate so much that the Bourdon tube inside it can crack, releasing media."

One way to prevent vibration-related problems is to install liquid-filled gauges, which help to dampen the vibration and also protect the internal components. The liquid case filling also lubricates the internals and prevents corrosion.

2 - Pulsation.

When media rapidly cycles through the pressure system of the gauge, pulsation occurs. If the pointer pulsation increments are greater than 5 percent of the full scale value, you must intervene to prevent damage to the gauge.

A restrictor or throttle screw can solve most pulsation problems. It will slow down the media by reducing the size of the orifice and averaging out pressure fluctuations. For extreme pulsation, such as spiking and surging, a snubber works best. It allows more control over the amount of snubbing and the flow of media into the pressure system.

3 - Temperature:

If a gauge is exposed to a media with very high temperature, the metal alloy in the gauge may not be able to handle the temperature, causing the Bourdon tube to fail and release media.

Long pipes that separate the pressure gauge from the process, or small cooling elements with radiator fins, protect the pressure instrument from high process temperatures by allowing ambient air to flow over the surface area. In environments with steam, using a siphon with the pressure gauge cools the steam down and drives it back to its liquid state before it gets to the gauge. With very high-temperature media, a diaphragm seal with a cooling element will isolate and prevent the hot media from reaching the gauge.

4 - Spikes or Overpressure.

In some cases, process media may be under extreme pressure. Pumps also can cause spikes and over-pressure, both of which shorten gauge life. Gauges can fail as a result of these pressure spikes. When workers turn pumps on or off, or open or close valves, a surge of media or oil goes through the pipe and hits the pressure gauge, which causes a huge spike. For example, if a line is pumping 100 psi of oil, someone may install a 100 psi gauge, not realizing that this leaves no allowance for any overpressure, surging, or spiking. When pressure does increase, it causes overpressure on the gauge, sometimes to the point of where the Bourdon tube ruptures and leaks media.

One way to prevent this problem is to use a different type of gauge, such as 200 psi gauge for a 100 psi line. Another option is to accessorize the gauge with an over-pressure protector, which would shut off the pressure if there were a surge or spike. Overpressure protectors are designed to avert spikes and the resulting overpressure damage. These protectors provide protection for pressure gauges against the effects of pressures exceeding their maximum pressure ratings.

5 - Corrosion.

Corrosion problems can occur with gauges if the process media is not compatible with the pressure gauges' wetted parts material. Media can sometimes be so corrosive that it corrodes through the Bourdon tube and leaks onto the ground.

When the media is not compatible with the wetted parts material of the pressure gauge, a diaphragm seal is required to isolate the gauge from the media. The benefit of a diaphragm seal is that it can be made of special alloys that are designed to withstand corrosion. It is essential to select a diaphragm material that is compatible to the process media.

6 - Clogging.

In some processes, the media may eventually crystallize or may otherwise harden. If a standard gauge is used in such an application, it will eventually get clogged and will not be able to read the pressure accurately. For example, if a gauge gets clogged and ends up reading zero pressure, when in fact it is under tremendous pressure, there will be an unexpected and dangerous release of media if and when the piping system is opened. Diaphragm seals protect the pressure gauge from solids, particles and viscous media by creating a barrier between the process and the pressure inlet of the gauge. There is also an option to equip the diaphragm seal with flushing ports to constantly flush material out from the diaphragm surface.

7 - Mishandling and Misapplication.

Gauges can also fail prematurely if they are misapplied. For example, a gauge may have been ideal for the application when it was installed. However, if there are equipment modifications or more components subsequently added, this can change process parameters, resulting in the gauge not longer being right for that application. Misapplication can lead to serious problems associated with failure and subsequent media release.

It is also possible to accidentally damage gauges during installation. Gauges should be installed using the proper sized wrench on the flats. Gauges should never be installed by turning the case. After installation, gauges can also fail if they are mishandled or used improperly.

Why Problems Can Occur With Gauges

In a typical process plant, about 25 percent of pressure gauges have failed or are about to fail.[1]
A lack of management focus and worker education and experience are key reasons plants experience issues with gauges.

Management Focus.

Gauges are often viewed as inexpensive commodities, with management failing to realize that they are very critical parts of the processes. Often, plants will replace failed gauges with the same replacement gauges, which will also fail, only to have to be replaced again. Eventually, their luck may run out and a serious failure may occur because the incorrect gauge has been installed.

Why do gauges tend to receive such little attention from management? Management tends to focus on the expensive equipment - $100,000 or million dollars items. They often don't understand the importance of focusing attention on $100 gauges. This is not the fault of management. The problem is that no one has explained to them just how important gauges are.

As noted earlier, though, ignoring instrumentation can cause serious safety and financial problems. Plant process gauges can be thought of in the same way as vehicle gauges. If the temperature gauge on an automobile is defective, it is possible to destroy the engine as a result of overheating that is not registered by the defective temperature gauge. Similarly, if the oil pressure gauge on an automobile is defective, it is again possible to destroy the engine as a result of seized components resulting from loss of oil that is not indicated by the defective oil pressure gauge.

Worker Education and Experience.

A lack of worker education and experience also contributes to gauge issues. In many instances, when maintenance technicians look at gauges, they don't pay much attention to them unless they leaking or look as if they are going to explode. They don't realize all of the information that that gauge can provide. In addition, technicians may not realize that a gauge has failed or is about to fail. Again, as long as it is not leaking media, they will often just keep walking. This is rarely the fault of the technicians. Many have not been properly trained to understand process pressure gauge performance.

Along the same lines, the industry has been experiencing a "brain drain" for several years, with large numbers of experienced technical workers and engineers retiring and being replaced by younger people who lack the education and experience to understand what gauges are telling them, or, worse, not being replaced at all.

According to research conducted by Cambridge Energy Research Associates in 2007, more than half of all oil-field professionals will reach retirement age during the next decade. The Interstate Oil & Gas Compact Commission found that the average age of production and service company employees is 46 to 49 years, and the average retirement age in the oil and gas industry is 55 years. [2]

Historically, plants had specialists whose jobs focused exclusively on instrumentation. They spec'd the gauges, maintained them, and made sure they were operating correctly. However, as plants got leaner in recent years, they started letting these people go. Plants first shifted gauge responsibility to the pipefitter whose main responsibility was to make sure the pipes were put together correctly and not leaking.

[1]WIKA data aggregated from more than 250 instrument audits.
[2]Peter Peek, Jocelyn Fenard, Patrick Gantes, Christine Theiler. December 2008, "Skills Shortage in the Global Oil and Gas Industry: How to close the gap" Part I. Pages 14-15.

Then, as plants began letting pipefitters go, they shifted the responsibility to the unit operators or technicians, who now install the gauges. These people usually don't have any experience or knowledge of how to install instrumentation. They usually just replace old gauges with similar gauges. The problem is that the gauges are usually the wrong ones to begin with. The result is that, these days, few plants have people who have much if any education and experience with instrumentation.

Taking Stock of Process Pressure Gauges

When it comes to improving the safety, productivity and profitability of process plants through an effective gauge management program, companies can take a couple of steps. First, an understanding of the vital importance of gauges must be established. Second, workers must understand what kind of gauges will be effective.

Focus on Gauges:

World-class process plants maintain their assets proactively. In such plants, pressure gauges receive as much attention as pumps and control valves. Management understands the importance of proactively maintaining their assets, and they realize that one of the best ways to maintain their most expensive assets is to properly source and maintain the gauges that protect these assets. That is, they understand the value of spending $100 for a gauge that can subsequently prevent damage or destruction to a $100,000 or $500,000 piece of equipment and the resulting negative safety, downtime and regulatory implications.

There is an interest in predictive maintenance these days in process plants, which is a good thing. A lot of this, though, involves expensive software. In addition, when something is out of kilter, the system may be set up such that technicians may have to wait for it to get far enough out of the normal range before it will generate an alert on the computer. However, properly-maintained pressure gauges, studied by people who know what to look for, can be very inexpensive but reliable and useful mechanical predictive maintenance tools.

Gauge Procurement Considerations:

Over time, as noted earlier, the status of pressure gauges has unfortunately diminished to that of commodities in many process plants. In order to ensure process plant safety, productivity and profitability, it is essential that plant professionals understand the importance of selecting the right gauges for applications.

It is also important that these be high-quality gauges, not "commodity" gauges. That is, not all process gauges are created equal. The so-called "lowest price" gauges appear attractive at first, but are rarely the "lowest cost" in the long run, because they usually have to be replaced more often, leading to excessive downtime and inconvenience. Worse, they may fail unexpectedly, leading to catastrophic results.

Consider these 10 steps to select the right gauges that will help to ensure safe, productive and profitable plant processes.

1 - Know what is in the pipe. This way, you can select a gauge alloy that will be compatible with the process media.

2 - Next, know the temperature of the media. If it is very hot, for example, it will be unsafe to use a standard pressure gauge. A long pipe, siphon or diaphragm seal assembly may be needed to cool the media sufficiently before it reaches the gauge internals.

3 – Third, know the ambient conditions surrounding the gauge. For example, if it is corrosive around the gauge, you need to make sure that the external components of the gauge such as the case and window can handle the environmental conditions.

4 - Next, know the pressure in the piping that the gauge is meant to measure. A best practice is using a gauge that is designed for two times the normal operating pressure. For example, if normal pressure is 100 psi, use a 200 psi gauge.

5 – Know the conditions that affect gauge wear. These could include mechanical vibration, pulsation, pressure spikes, etc. For example, if vibration exists, a gauge with a liquid-filled case is recommended. If there is pulsation, use a throttle plug or restrictor accessory.

6 - Select gauges that are designed for a particular service. Process type gauges are designed with strong Bourdon tubes and other heavy duty components. If a gauge is misapplied, the internals of the gauge may not be rugged enough to resist failure and subsequent media release in these applications.

7 - The gauge case must be rugged and have the correct safety features. In some case designs, if there is a rupture of the Bourdon tube, media will be released out the front of the gauge, which can spray workers. Select a gauge where media would release out from the back of the gauge.

8 - Accuracy is important. In most process plants, +/-0.5% accuracy is the norm, and +/-1% is acceptable.

9 - Size is also important. The gauge dial needs to be large enough that operators can read it from a distance – four inches or larger.

10 - Finally, the gauge should be mounted properly. A gauge is usually mounted by its stem, but it can also be mounted in a panel or against a flat surface.

Looking Outside for Assistance

As noted earlier, it can be difficult for plant management to devote enough time and attention to an effective process pressure gauge program. Similarly, it can be difficult for technical personnel to get up-to-speed in terms of education and experience related to process pressure gauges.

There has been a growing trend over the last 30 years in virtually all industries for companies to outsource services that are not considered "core competencies." That is, companies have come to realize that it makes sense to focus their attention on their missions and what they do best, and leave secondary and peripheral activities to third parties that specialize in these other activities and services. Becoming experts in pressure gauges is certainly not a core competency for refining plants, petrochemical, and chemical plants, so this is an area that can easily be outsourced to an experienced third-party audit team.

If a plant elects not to gain the experience that it needs on its own, free assistance is available. To ensure that gauges are working properly and are able to protect the valuable assets that they were designed to protect, a five-phase instrument review is recommended. A best practices review identifies gauges with unwanted conditions and recommends corrective actions to ensure optimum performance. The review identifies issues that affect reliability, longevity and safety. Such a review can uncover evidence of pulsation, vibration, pressure spikes, excess process or ambient temperature, installation damage, clogging, chemical incompatibility, and other problems that can lead to serious problems.

Phase 1 - Interview. This involves discussions between the third-party audit team and plant personnel. During the interview, plant personnel learn the purpose of the instrument audit, how it will be conducted, and how they will benefit from the audit. The interview also helps determine the best way to match the audit to the needs of that specific plant.

Phase 2 - Pre-Meeting. This involves determining the units or sections of the plant to audit, as well as outline roles and responsibilities. It also identifies prohibited and exempt areas of the plant. The third-party team reviews problematic instrument installations and discusses safety procedures with plant personnel.

Phase 3 - Instrument Application Audit. This involves having key plant personnel lead the third-party team on an actual plant walk-through. The audit team reviews the target gauge installations, identifies gauges that demonstrate unwanted conditions, as well as those that demonstrate installation best practices.

Phase 4 - Post-Meeting. This involves explaining the instrument audit results to key plant personnel, specifically noting gauge installations that pose immediate safety concerns. Together, the audit team and plant personnel determine which installations to investigate immediately and which to block out. The audit team and plant personnel then set schedules for the follow-up meeting.

Phase 5 - Follow-up Meeting and Presentation. The third-party team presents its general findings and the executive summary resulting from the instrument audit. The team also presents a spreadsheet that lists details about the gauges from the instrument review, color-coded by severity. The team also provides recommendations for improvements based on experience and best practices. Together, the audit team and plant personnel develop a strategy for moving forward.

The plant should also consider leveraging the audit team in other ways. For example, over time a plant's storeroom inventory can get out of hand. A qualified third-party audit team can help plant personnel rationalize, consolidate and manage their inventory of pressure gauges. For example, if a plant stocks spares for 10 or more ranges, the audit team may be able to help reduce that number to five or six ranges.

Essentially, when used properly, gauges play a significant role in plant safety, productivity and profitability. They act like silent alarms, indicating when something is seriously wrong or helping to diagnose deeper issues within systems and processes that can eventually lead to serious safety issues. To ensure safety and reliability at your plant, make certain the proper gauges are in place by reviewing your processes, gauges and your inventory. Everything is at stake.

REFERENCES

1. WIKA data aggregated from more than 250 instrument audits.

2. Peter Peek, Jocelyn Fenard, Patrick Gantes, Christine Theiler. December 2008, "Skills Shortage in the Global Oil and Gas Industry: How to close the gap" Part I. Pages 14-15.

ExxonMobil
Research and Engineering

Best Practices in Pilot Plant Piping

Richard Palluzi
April, 2013

The ideal piping for a pilot plant would be:

- Leak resistant
- Easily modified
- Able to accommodate varying requirements
- Easy to install
- Low cost
- Available in many different materials of construction
- Available in all sizes
- Have fittings available in numerous configurations

Leak Resistance

- Pilot Plants are more subject to leakage than process plants
  - Smaller size
  - Larger number of joints
  - Constant modifications
  - Need to accommodate accessibility, flexibility and change
- Common types of piping used in pilot plants
  - Piping
    - + Threaded
    - + Welded and flanged
    - + Groved
  - Tubing
    - + Compression fittings
    - + Vacuum fittings
    - + Mechanically joined

3

ExxonMobil
Research and Engineering

Best Practice: Tubing

- Available in numerous materials
- Fittings are readily available up to 1" OD
- Easier to install and modify
- Requires less skill, tools and training for installation
- Smaller and lighter than pipe
- Lower cost than pipe
- Ability to bend
  - Reduces need for fittings (lower cost and less potential for leakage)
  - Easier to install in tighter locations
  - Allows for easier removal of components for cleaning, modification or inspection

4

ExxonMobil
Research and Engineering

Best Practice: Compression Fittings

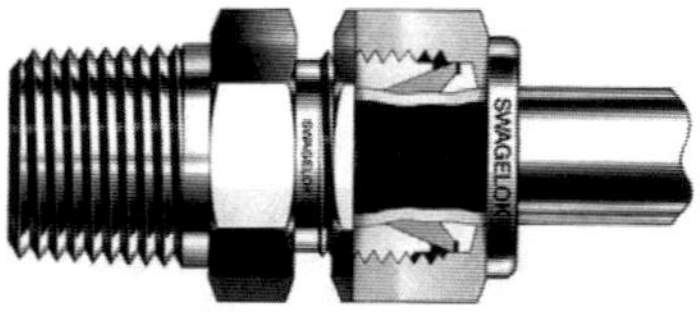

- Better leak resistance than threaded fittings
- Requires few tools and less training
- More expensive than pipe fittings
  - Significantly faster to install so labor savings usually make then cheaper overall
- Each fitting is a potential disassembly point
  - Do require clearance although much less than piping

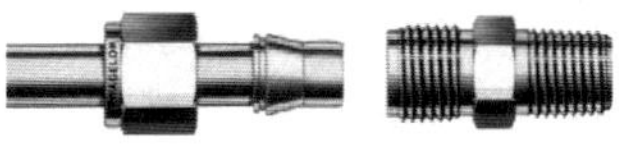

- Significantly better make/break life than pipe fittings
  - But not infinite!

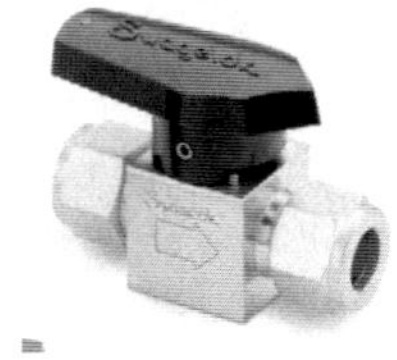

- Numerous specialty fittings available
- Many small valves and associated components available
- Most compression fittings have limitations on tubing hardness
  - Confirm the tubing you ordered is within specification

5

ExxonMobil
Research and Engineering

Best Practice: Compression Fittings

- Recognize that compression fittings larger than ¾" may require mechanical assistance
- Compression fittings have both minimum and maximum wall thicknesses
  - Too thin and the tube may collapse without swaging
  - Too thick and the tube will not deform enough to prevent blowout
- Drilled through compression fittings, used for dip tubes, thermocouples and similar uses, lose significant pressure ratings

OD(inches)	Typical Pressure Derating
½ and less	25%
5/8-3/4	50%
7/8 - 1	75%

6

ExxonMobil
Research and Engineering

Best Practice: Vacuum Fittings

- Significantly better leak performance than threaded fittings
  - Usually requires some welding to avoid additional joints
- More expensive than compression and pipe fittings
- Each fitting is a potential disassembly point
- Zero clearance fittings
- Long term make/break life
- Gasket surface is more prone to damage
  - Awareness can reduce frequency
- Decision to use O-ring style or metal gasket is usually based on temperature and pressure

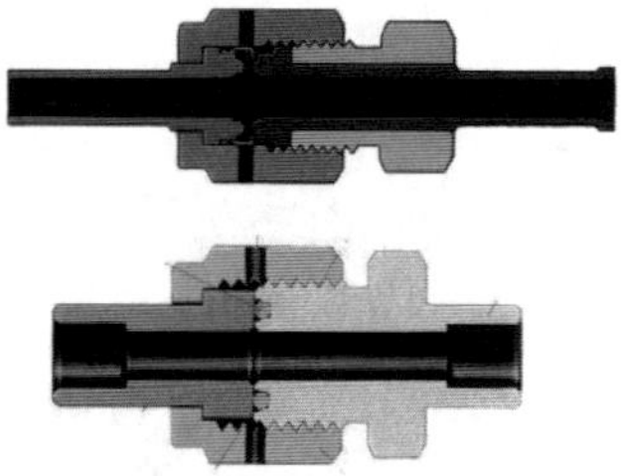

Best Practice: Threaded Piping

- Make sure the die is sharp and in good condition or the threads are likely to be rounded and more prone to leakage
- Make sure the threading is the right length
  - Longer or shorter lengths will leak
- Use good quality fittings
  - Particularly unions
- Avoid threaded connections in cyclic service
  - Temperature
  - Pressure
- Lightly coat unions with lubricating oil before each assembly
  - The coating greatly helps it to seal in gas service
- Recognize that threads over 1" NPS are increasingly difficult to seal
  - Particularly in gas service

Best Practices: Flanged Pipe

- Use raised face flanges whenever possible as they require less force to seal
- Avoid threaded flanges
  - Adds another potential leak point
  - Often makes bolt alignment difficult if retightened
- Tubing is usually a better approach for pipe sizes ½" and smaller

Best Practice: Specialty Piping

- Understand when specialty piping may be a viable or more effective option
- Grooved piping allows for fast and easy disassembly

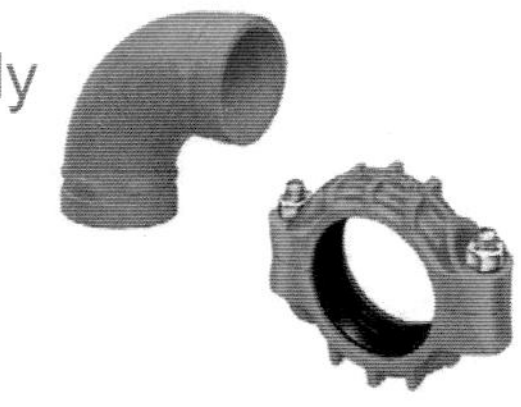

  - Lower pressure (1,000 psig is typical)
  - CS or SS only
  - High materials costs
  - Some specialty equipment (grooving tool) required
  - Sizes below 1" NPS very limited
- Sanitary fittings allow for ease of cleaning

  - Available for piping or tubing
  - 304 or 316 SS only
  - Lower pressure (300 psig is typical)
  - High materials costs
  - Sizes below 1/2" NPS very limited

Best Practice: Specialty Piping

- Flare compression fittings are excellent for thermal cycling but are generally only available in CS and brass
  - Good for refrigeration systems
- Longer ferrule designs and cone and collar type compression fittings extend the range of compression fittings to 15,000 or 60,000 psig respectively
  - Much more expensive

11

ExxonMobil
Research and Engineering

Best Practice: Mounting

- Use vacuum fittings where frequent removal or cleaning is required
  - Zero clearance makes layout easier
  - Ability to replace gaskets maximizes service life

- Mount the valves not the tubing wherever possible
  - Minimizes the stress on the fittings created when operating the valves
  - Will reduce long term leakage

- Avoid unsupported elements
  - Long spans
  - Heavier components
  - Flimsy supports
  - All lead to increased leakage
- Use bends to allow for thermal expansion
  - Even small temperature swings promote leaks over time

12

ExxonMobil
Research and Engineering

Best Practice: Reduce Joints

- Fewer joints always mean fewer potential leak points
- Use specialty fittings to minimize joints
  - Long term savings outweigh costs over time
  - Requires some planning or stocking
- Weld joints that are:
  - In difficult to reach locations
  - Not likely to need to be opened
  - Are on systems subject to routine or frequent temperature swings
  - Under insulation

13

ExxonMobil
Research and Engineering

Best Practice: Remember Maintenance

- Almost everything eventually needs maintenance
- Make sure easy removal is provided for:
  - Instruments
  - Rotating equipment
  - Filters and strainers that need to be removed to be changed
  - Reactors
  - Regulators
  - Relief devices
  - Control valves
- Decide if you have clearance or need zero clearance fittings

14

ExxonMobil
Research and Engineering

Best Practice: Leave Room

- Pilot plants are always modified
- Building a unit in too small a space:
  - Adds to construction costs
  - Makes future changes difficult or impossible
  - Greatly increases maintenance costs
- The Golden Rule: Plan on at least twice the largest space you think you'll need
- Piping runs eat up space
- Insulation easts up space

15

ExxonMobil
Research and Engineering

Best Practice: Training

- Train your operators and craft support regarding best piping practices
- Always use a holdback whenever making or breaking any fitting
  - Otherwise the fittings o neither side will start to leak
- Make up the fitting the proper amount
  - Not too much, not too little
  - Consider gauge blocks for compression fittings
- Never remove a support
- Make sure any changes or modifications are properly supported

16

ExxonMobil
Research and Engineering

Best Practice: You Get What You Pay For

- Buy good quality pipe, tubing and fittings
- Use high quality gaskets and sealants
  - Teflon tape and pastes generally work better than dopes
  - Pastes require a minimum time for hardening
  - Silver plated vacuum gaskets seal better than plain stainless
  - Spiral re-enforced gaskets seal easier and last longer than flat gaskets
  - Paying more for a good quality seal always ends up being more cost effective!

17

ExxonMobil
Research and Engineering

Best Practice: Heat Tracing

- Avoid joints wherever possible
- Use bends to accommodate thermal expansion
- Use standoffs to avoid having supports becoming heat sinks
- Use vacuum fittings when joints are necessary
- Mark insulation for joint location for ease of maintenance

18

ExxonMobil
Research and Engineering

(54b) Successful Operation of a 1 BPD Fischer-Tropsch Pilot Plant

Juan Inga and **Lindsay Hernandez**, SGC Energia, Houston, TX

In 2012 SGC Energia commissioned, started up, and operated a pilot scale Fischer-Tropsch process for the demonstration step of the XTLH solution in Pasadena, TX. This plant was used to confirm design plans for future larger scale projects as well as to gather additional data. Running a Fischer-Tropsch reactor at a nominal capacity of 1 BPD, SGC Energia gained valuable information from lessons learned and the challenges of the start-up process. Upon the conclusion of the successful run, this moveable skid is being relocated to provide future operator training for a commercial unit. This commercial plant will be located in the Gulf area.

SGC Energia has operated different FT reactor technologies at a 1BPD level, plus others on a smaller scale. Commercial sized samples have been produced from these operations. Currently the SGCE Fischer-Tropsch reactor choice greatly reduces the risk involved, making the system more robust; and therefore more equipped for commercialization.

(54c) Continuous Improvement in Pilot Plants for Safer Operation

Kiran Gupte, BP Refining & Logistics Technology, Naperville, IL

Continuous Improvement in Pilot Plants

for Safer Operation

Kiran Gupte

BP Refining & Logistics Technology

150 W. Warrenville Rd

Naperville, IL 60563

BP's Refining & Logistics Technology group has successfully designed, commissioned and operated pilot plants in support of refining processes over the last 20 years. Of course, over the years the management practices have evolved with newer methodologies. The "Continuous Improvement Process" is important to critically review all operations to further strengthen the safety in design and operation. In addition to the old and reliable "Process Safety Review (PSR)" process, the newer "Management of Change (MOC)" procedures play the most critical role. An "Inherently Safer Design" is not only for a new pilot plant during the conceptual design phase but it should also be considered in every improvement. The "Continuous Improvement" process helps to bring out and consider smaller steps in improving operation of the pilot plants. The cumulative effect of every improvement step however can never be over emphasized. This presentation will highlight the challenges of operating small scale pilot plant and methods to continually improve these units for safer operation. As the pilot units get used for ever expanding types of projects, the attention to "Operating within the Design Envelope" is important for safe operation with zero incidents. All these aspects are an integral part of the overarching Operations Management System. The safe practices developed are applicable to all types of equipment including small-scale catalyst screening reactors, small-medium autoclave reactors, distillation apparatus, and large pilot plant and demonstration reactors handling a broad range of hydrocarbon feedstocks.

(54d) Using Dimensional Analysis to Design Pilot Plant Experimental Programs

Jonathan H. Worstell, Consultant, Richmond, TX

Pilot plants are capital intensive and expensive to operate. The most expensive item of a pilot plant budget is its experimental program. A poorly designed pilot plant experimental program requires more operating campaigns than originally budgeted. Thus, such a pilot plant will "go beyond" its budget --- a non-optimal situation.

This presentation discusses using Dimensional Analysis as a method for designing pilot plant experimental programs. First, Dimensional Analysis helps identify the major parameters impacting the chemical process. Second, Dimensional Analysis shows how to minimize the experimental program for a given pilot plant, thereby reducing its total operating expense.

Dimensional Analysis is not new: it has just fell out of favor with chemical engineers. It is time to rethink its use.

Design of Batch Scheduling

Aaron Herrick, Chemstations, Inc.

Chemstations™ www.chemstations.com

Abstract

Design and scale-up of new batch processes presents a unique set of challenges versus design of continuous processes. One key difference is optimization of the time required for each process step. Rules of thumb are often used due to lack of data or tools.

Improvements in dynamic simulation software have allowed for more rigorous modeling of increasingly complex systems and scenarios. Modeling of batch process steps in the design phase can highlight potential problem areas such as heat balance limitations, short residence times, reversal of flows, and pressure/flow dependencies. This allows for efficient optimization of step scheduling using the latest dynamic simulation techniques. Examples of batch designs are presented using CHEMCAD simulation software. A new method is included for simplification of handling any number of process steps and control elements.

Chemstations™ 1

Agenda

- Current practices in design of batch scheduling
- Improvements made possible with
  - Rigorous modeling methodology
  - Microsoft Excel™ to drive a schedule in a simulator
- Example using CITRON
- Example using WINTEK
- Summary & suggested procedure

Chemstations™ 2

Current practices in design of batch scheduling

- Methods
  - Best guess & experience
  - Lab scale & pilot scale testing
  - Use a simulation
- Pros
  - Fast
  - Fairly easy
  - Gives a basic time and heat duty analysis
- Cons
  - Suboptimal schedule times
  - Might not account for equipment/process limitations
  - Difficult to do safety analysis until after process is running

Chemstations™ 3

CITRON (current practice)

Recover 120 kgs of 99.9999% pure Citronellal from 170 kgs essential oil using a 500L pot still with a 10ft packed column.

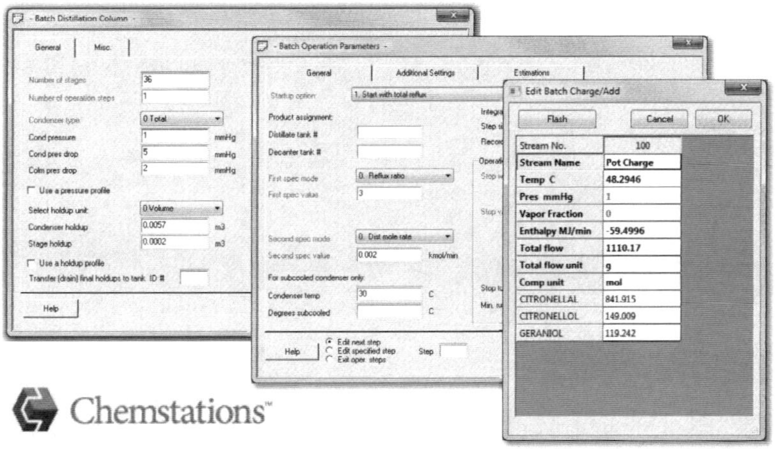

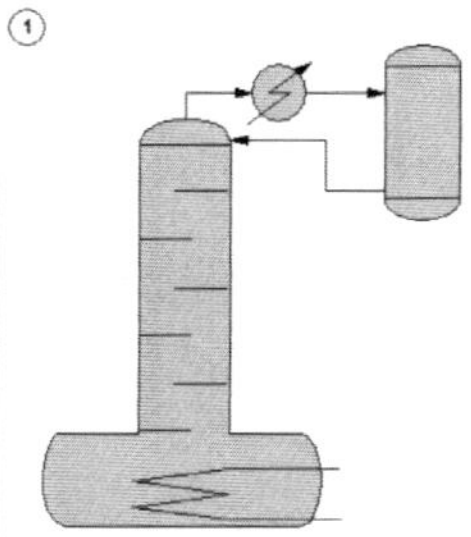

Chemstations™

4

CITRON (current practice)

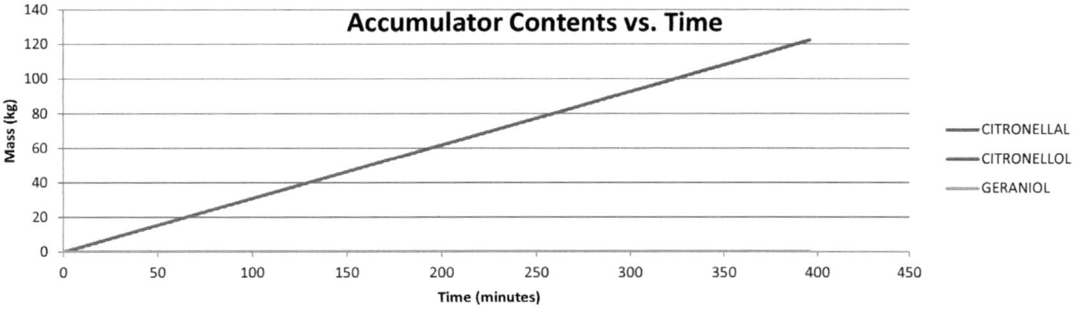

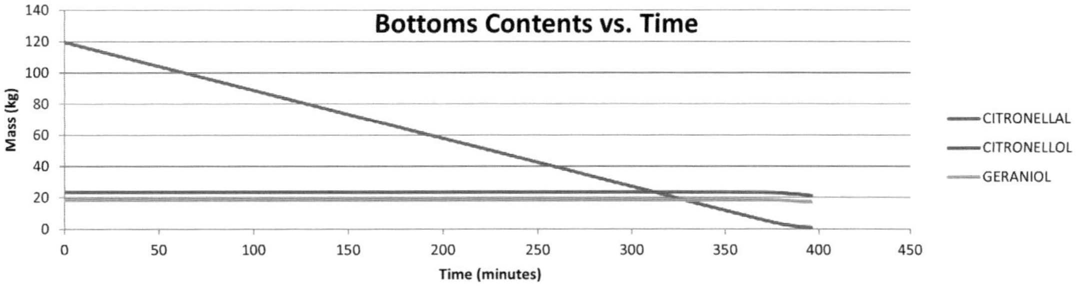

Chemstations™

5

CITRON (current practice)

What have we learned from the model?

- Heat duty/flow/time relationship
- Basic operation steps/timing
- Limited equipment sizes and specifications:
  - Heating requirements
  - Condenser requirements
  - Column dimensions

Chemstations™ 6

CITRON Dynamic (advanced practice)

- Expand our previous example to include
  - Rigorous heat exchanger geometry and performance
  - Event sequencing using DATAMAP to Microsoft Excel
  - Utilities modeling
  - Column metal heat transfer
  - Heat duty/ cooling water on control
  - Dry column startup to total reflux
  - Detailed engineering, e.g., nitrogen sweep on vessel, insulation, etc.

Chemstations™ 7

CITRON Dynamic

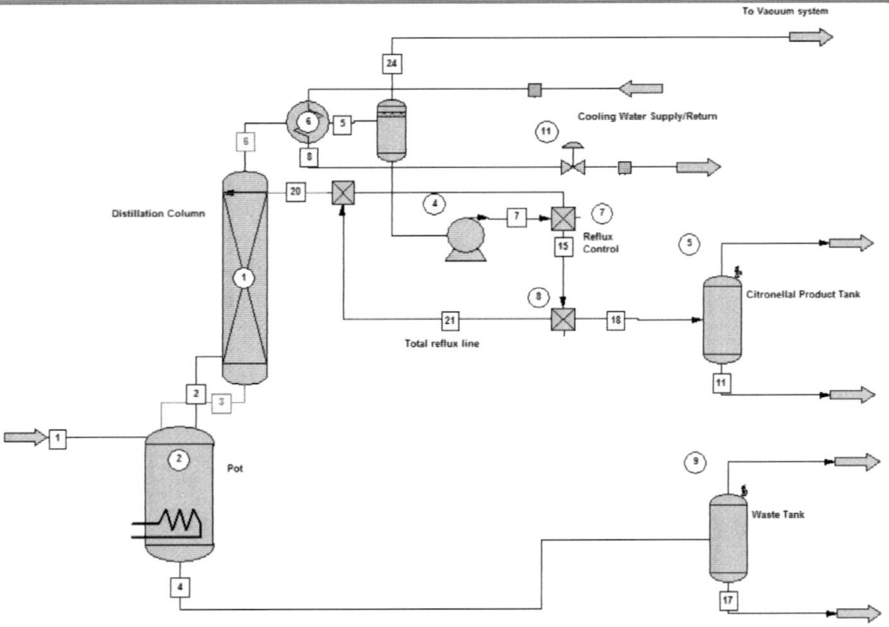

Chemstations™

8

Schedule of Events

- A schedule of events is made in Excel and must be connected to CHEMCAD:

Time(min)	CW Valve mode	CW Valve Opening	Total reflux	Product Tank	Waste Tank	G	H	Notes
0	4	0	1	0	0	36	0	Startup
21	4	0	1	0	0	29	0	Lower heat duty
25	4	5	1	0	0	29	0	Open CW Valve
50	4	5	0	1	0	29	0	Switch off Total reflux
473	4	5	[illegible]	[illegible]	[illegible]	0		Heater off, drain to tank
540	4	0	[illegible]	[illegible]	[illegible]	0		CW off

Schedule of heat duty to pot

Chemstations™

9

CHEMCAD to Excel connection

- CHEMCAD sends time information to Excel cell A2 via datamap
- Excel uses time, schedule, and VLOOKUP() function to determine values of process variables
- CHEMCAD collects the current time step data from row 2 of the Excel sheet and enters it into the flowsheet via datamap

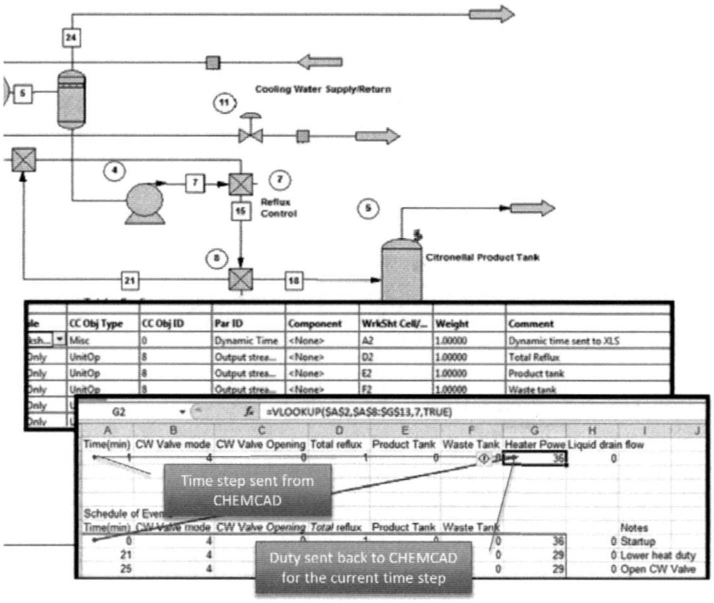

Chemstations™

10

CITRON Dynamic Operations

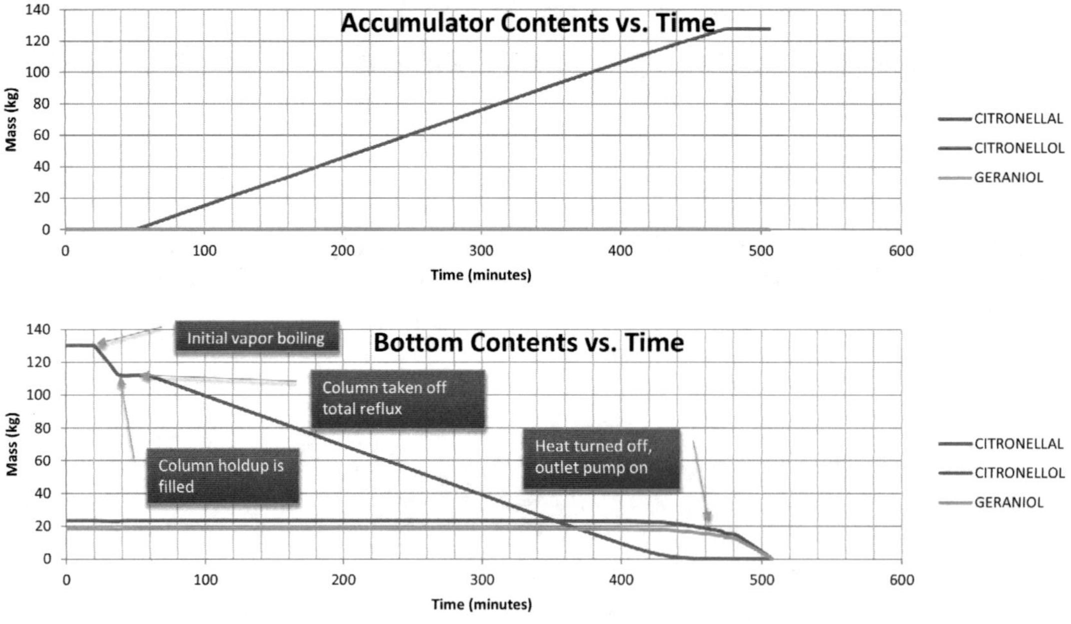

Chemstations™

11

Comparison of CC-BATCH and CC-Dynamics Column Bottom Flow

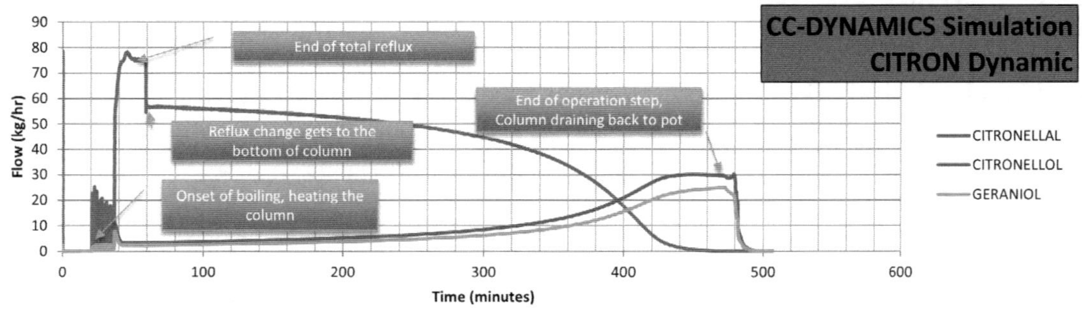

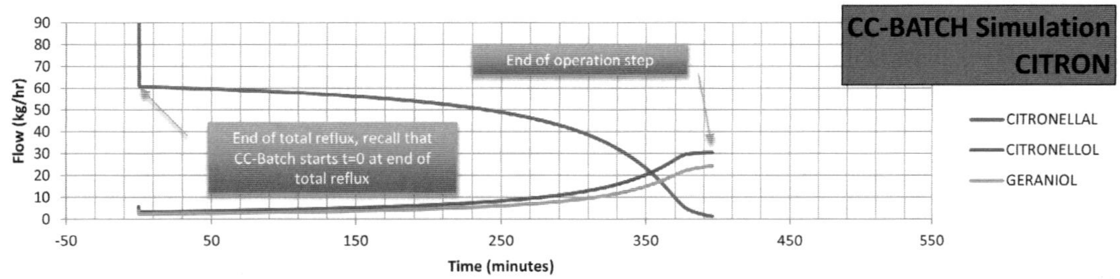

Chemstations™

12

CITRON Dynamic Conclusions

We learned from using dynamics:

- Detailed startup procedure
- Detailed operation steps/Sequence details
- Equipment performance limitations
- Vacuum load
- Utility demands
- Equipment optimization is now possible: checking condenser capacity, vacuum system capacity, column flood %, column insulation requirements, etc.
- Higher fidelity simulation provides higher fidelity economic calculation (campaign time and costs)

Chemstations™

13

WINTEK Batch Dehydration

Skid mounted solvent dehydration plant

Stripping water from a solvent stream in a two bed adsorber (mol sieve) system. One bed is active and one bed is regenerating under vacuum. Process was already built and operational before modeling analysis.

Process is scheduled with a defined sequence for opening and closing valves to allow one bed to regenerate while the other adsorbs. Rigorous equipment and piping models and pressure/flow calculation included (allows for reversible flow).

Rigorous simulation allows us to simulate effect of malfunction (RB3 blowdown valve malfunction)

Chemstations™ 14

WINTEK Process Flowsheet

2 Adsorption beds

2 PID controls

10 Actuated ball valves

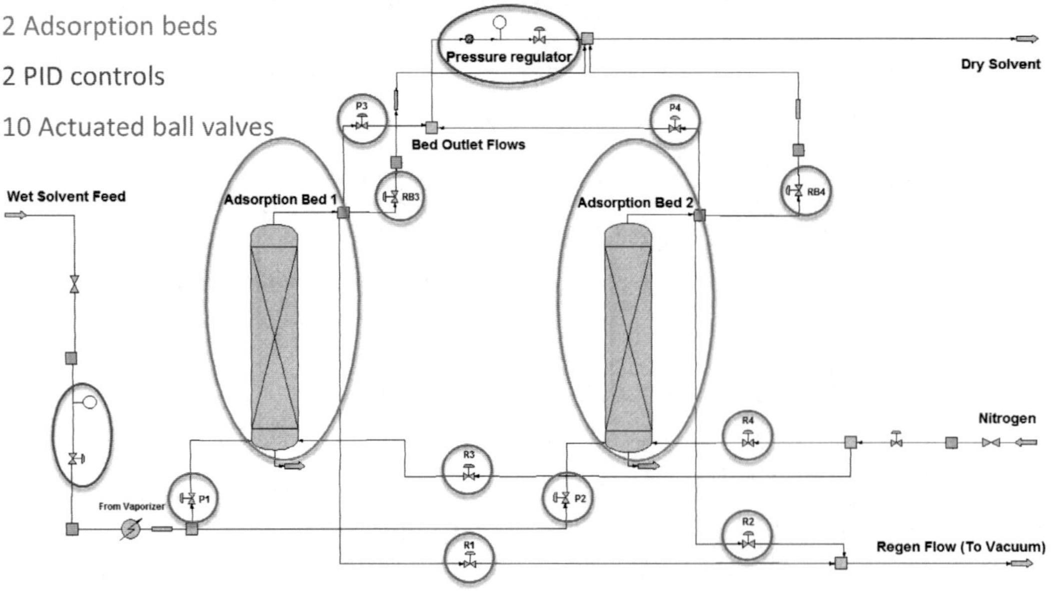

Chemstations™ 15

WINTEK Datamap and Schedule

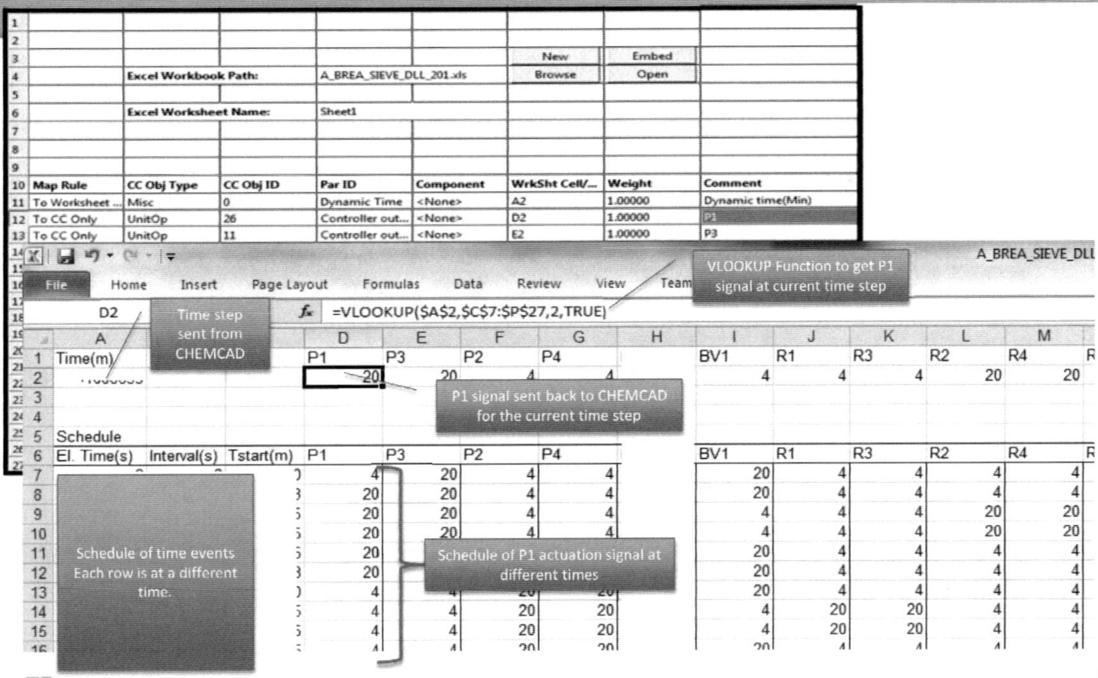

Chemstations™

16

WINTEK Process Flowsheet: Malfunction effect

RB3 is simulated blocked, preventing Bed 1 from blowing down properly. Effect is rigorously calculated.

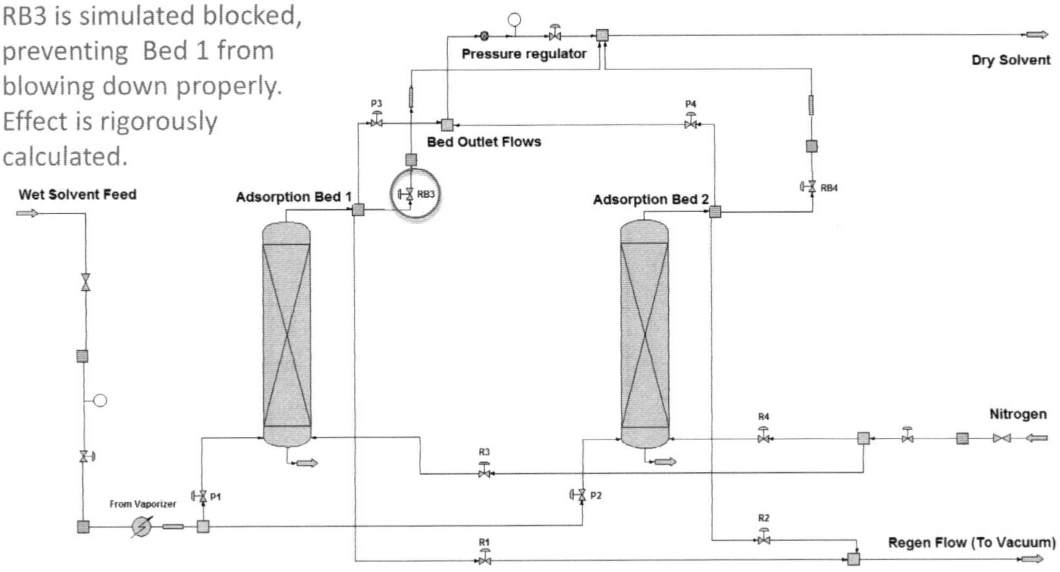

Chemstations™

17

WINTEK Adsorber Beds

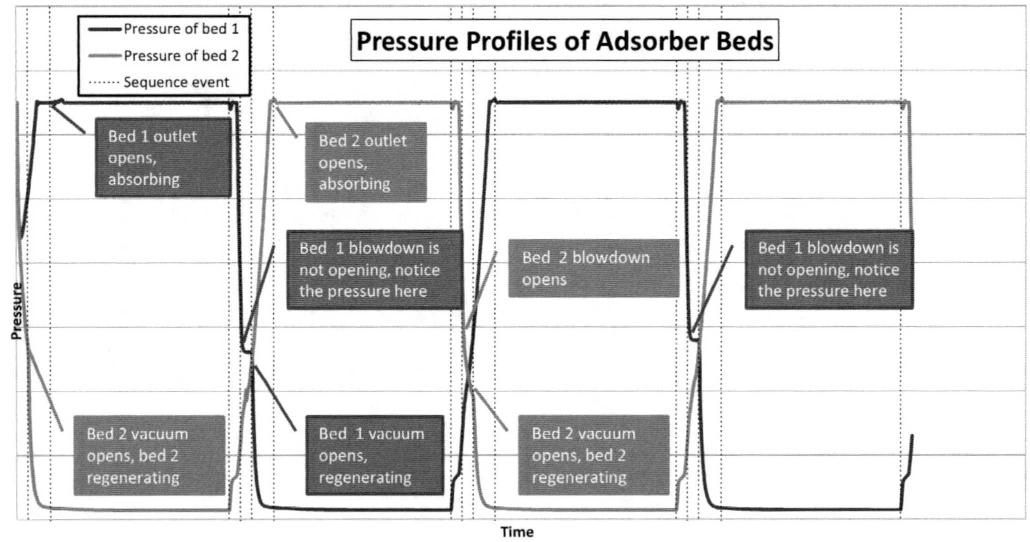

Chemstations™

18

WINTEK Anhydrous Solvent Flowrate

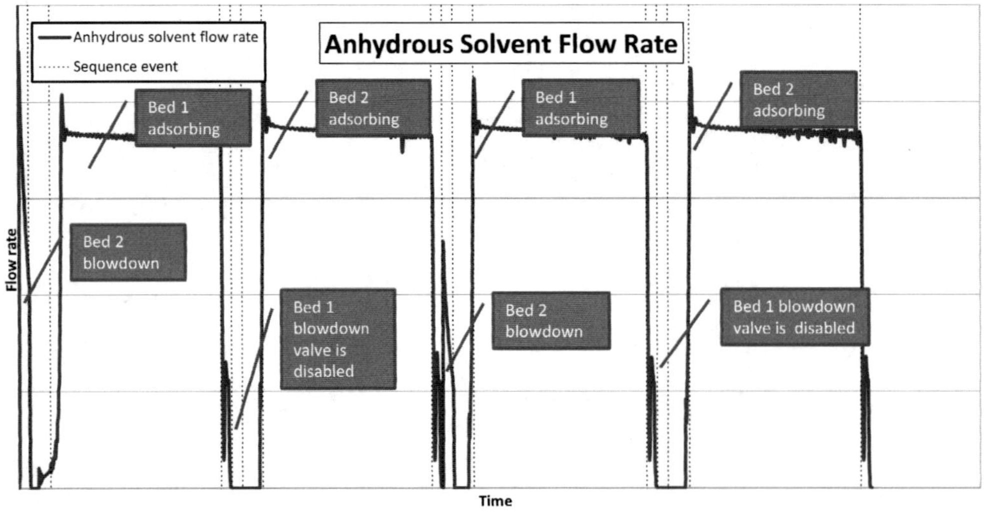

Chemstations™

19

WINTEK Process Flowsheet: reversing flow

P3 and P4 open, flow is forward through one valve, reverse through the other

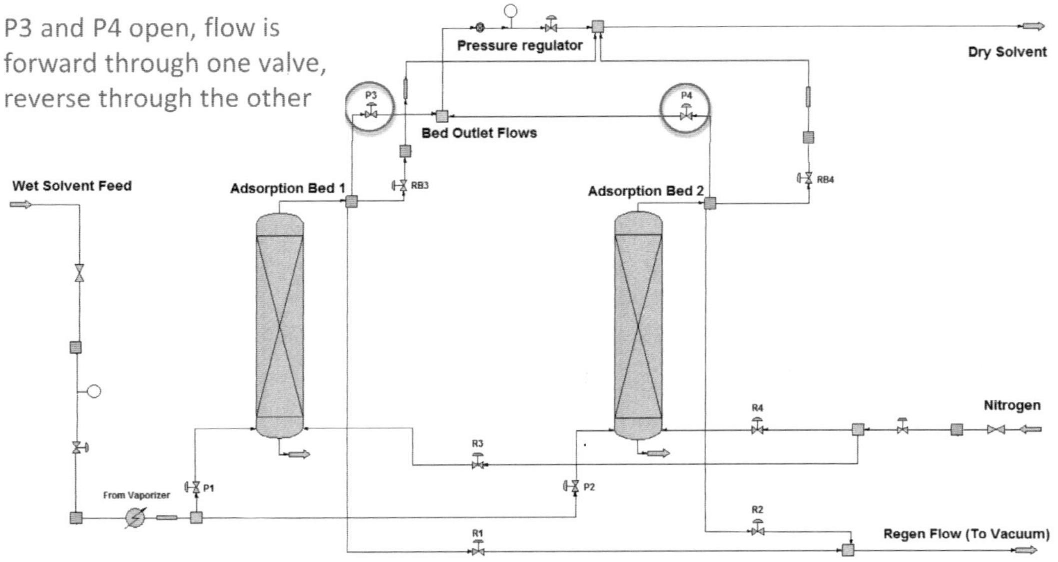

Chemstations™ 20

WINTEK Sequence of outlet and regeneration flows out of adsorber beds

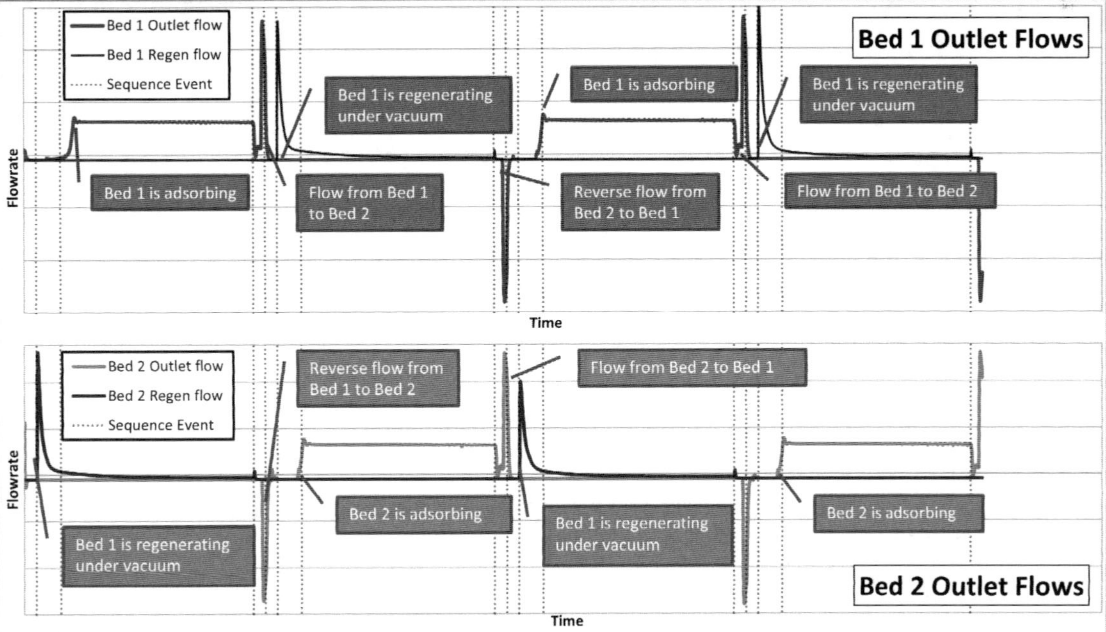

Chemstations™ 21

Optimization of schedule (future)

- By using an optimization engine (either the built in engine in CHEMCAD, or an external one connected to the Excel spreadsheet) the schedule of valve events can be optimized to maximize on-stream performance (blue area of chart).

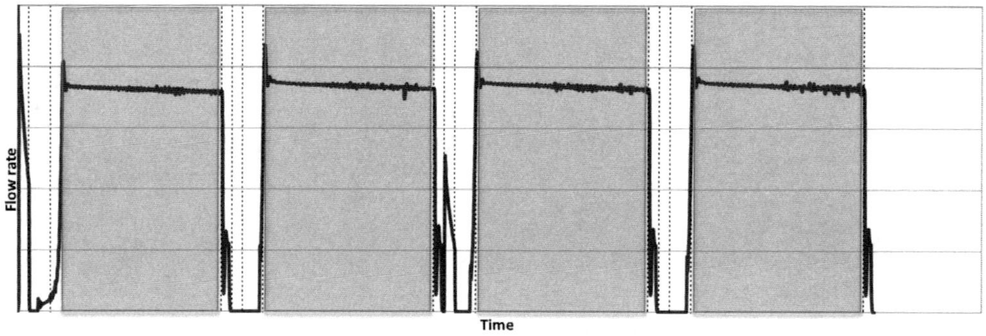

Chemstations™ 22

WINTEK Conclusions

What did we learn from our model?

- Discrete event scheduler with pressure and flow calculations (including reversible flow) allowed optimization of the process schedule
- Ability to identify and quantify
  - Bottlenecks
  - Sequence timing issues
  - Equipment limitations
  - Malfunction effects (bed 1 blowdown valve sticking)
  - Control loop tuning issues
- Ability to use the model for process scaling: process can be scaled up or down to meet customer requirements using the model before building/assembly
- Ability to optimize schedule of events

Chemstations™ 23

Suggested Procedure

1. Start with a simplified model (like CC-BATCH) to get heat duties, initial equipment sizes/specifications, and an initial timing/schedule
2. Build a rigorous dynamic model with as much detail as required (but no more than required) to solve the engineering problem (e.g., only add reversible flow if it is a concern)
3. Build an event sequence control scheme using DATAMAP and Excel starting with the information gathered in Step 1 above. Progressively improve the sequence by running the dynamic model with the scheduler

Chemstations™ 24

Special Thanks

Paul Winter, President and Owner

Wintek Corporation

http://www.wintek-corp.com

230 U.S. Highway 206 STE 401
Flanders, New Jersey 07836

Wintek has been providing equipment solutions to process applications operating at vacuum conditions since 1986, and serves the biodiesel, chemical, environmental, food, pharmaceutical, and plastics industries. Our focus is on process and equipment reliability for your application at a reasonable cost.

Chemstations™ 25

INLINE SYSTEM TO DETECT HYDRATES, WAX AND SAND

IMAGING Thomas M. Canty, P.E.
President, J.M. Canty Inc
Buffalo, NY 14094

ABSTRACT

Inline imaging systems are capable of viewing and digitally analyzing hydrates, waxes and sand in crude oil flow lines at pressure through 10,000 psi and temperatures up through 300 deg C. The technology relies on lenses that are made of fused glass which enable a quality, high speed image that detects and quantifies the size, shape, concentration and color of the suspended components in a crude oil stream. This allows for research and testing in addition to inline measurement and control to assure proper flow

INTRODUCTION

Vision systems are developed to the point where they can provide analysis and control in the oil and gas industry on a reliable and accurate basis. Design of hardware and software, and advancements in Ethernet communication technology have allowed vision based systems to see into process environments such as crude oil to detect and analyze particle constituents. Applications involving the monitoring and control of Produced Water concentrations, Water in Oil analysis, Multi-cut Sand, Oil and Water, Froth (Flotation Control) and Interface can also be effectively handled by vision with the benefit of visual optimization by the operator.

Vision analysis has been proven for many years in applications such as polystyrene bead manufacturing, sugar crystal growth, cell growth, abrasive manufacturing and many others, providing an extra dimension to process analytics that enables the process engineer to make evaluations based on his own expertise and then to use the information to better set software parameters that enable the system to more accurately analyze and control the process.

VISION TECHNOLOGY - HARDWARE

The conditions of the application require a vision system to have the ability to survive in an extreme environment, maintain explosion proof electrical ratings and be easily used by operators in order to be effective. The first point of discussion in this regard deals with the hardware mounted to the process.

Fused Glass Technology

The viewing mechanism of the vision system is most important and must be fitted with rugged, high pressure windows through which illumination and imagery can be conducted. The best option is the fused, glass to metal seal because it is rugged and can withstand pressure, temperature, shock and vibration. In addition, because it is fused directly to the metal probe it can be polished flush with the metal leaving no sites for particle adherence and contamination build up (see Figure 1). The fused glass interface (Fig. 2) is ground and polished to a flat, crevice free surface which remains clean during operation.

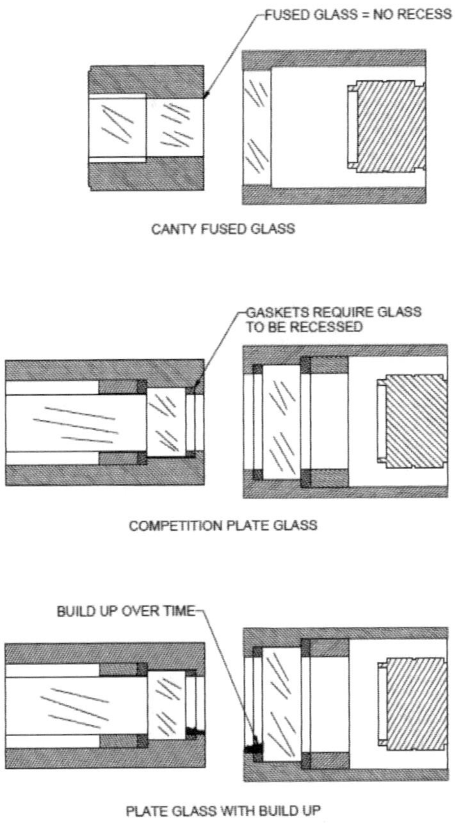

Fig. 1 – Fused Glass vs. Plate Glass Design and Cleanliness

Illumination

Illumination is critical to the success of vision based technology, and a consistent, high intensity source is the building block that allows software to define particles consistently and accurately. Another important feature of a complete system is the ability to control the measurement zone where the particles are illuminated and seen by the camera. The light pipe component shown in Figure 2 has adjustment capability in order that the gap between the camera and light pipe (measurement zone) can be adjusted to optimize the sensitivity of the system. Fluids can range from relatively clear (low turbidity) to quite heavy in particulate matter (high turbidity) and the control of the measurement zone enables the illumination to be controlled by means other than adjustment of intensity from the bulb. A typical light system is shown in Figure 3.

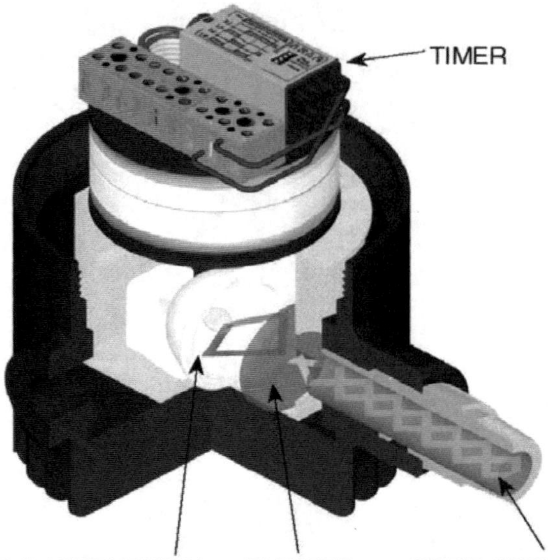

Figure 3. High Intensity Light Section

Cleanliness / Jet Spray Ring Technology

Development of a Jet Spray Ring generates a high pressure vortex rinsing action to remove tough deposits from Vision Systems. Originally designed for the U.S. Space Shuttle program, the Jet Spray Ring accepts a rinsing/cleaning solution and distributes the pressurized material through the specially designed rinse ports. These rinse ports have been engineered to provide maximum cleaning power in a compact package. The Jet cleaning action expels all debris from the Vision System glass, providing an exceptional view into the process. For petroleum applications the spray ports are designed into the instrument probe and maintain the configuration on the measurement zone.

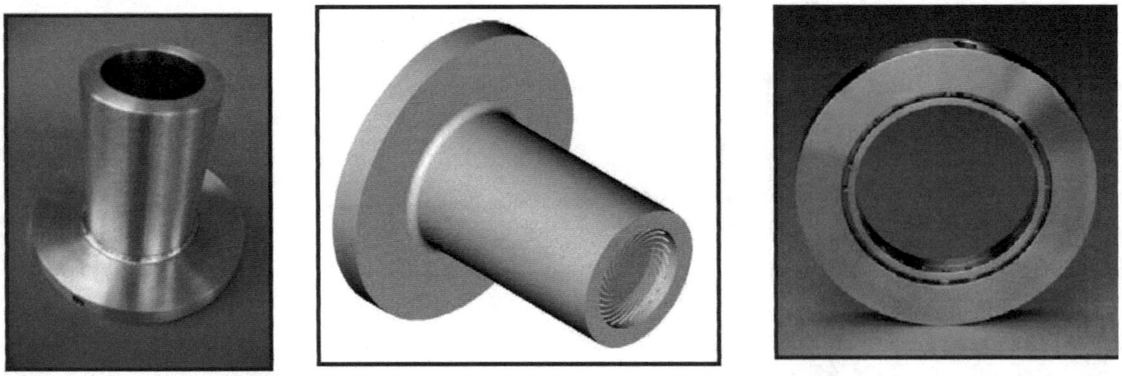

Fig. 4 – Typical Jet Spray Rings

Fig. 5 – Jet Spray Ring Cleaning Action Sequence

Typical Instrument Configurations

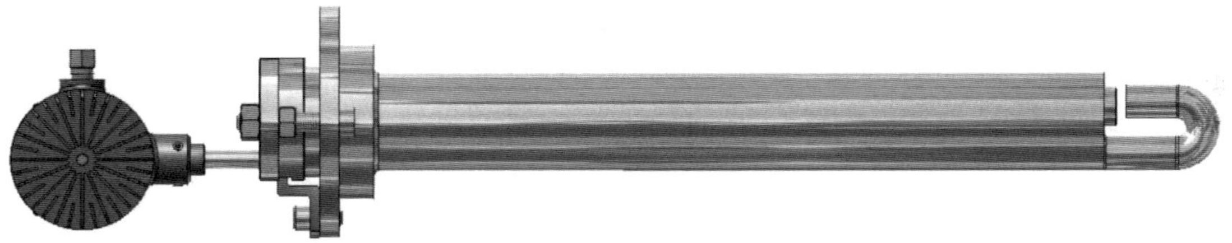

Fig. 6 - Inline System for 6" Flow Line and Larger

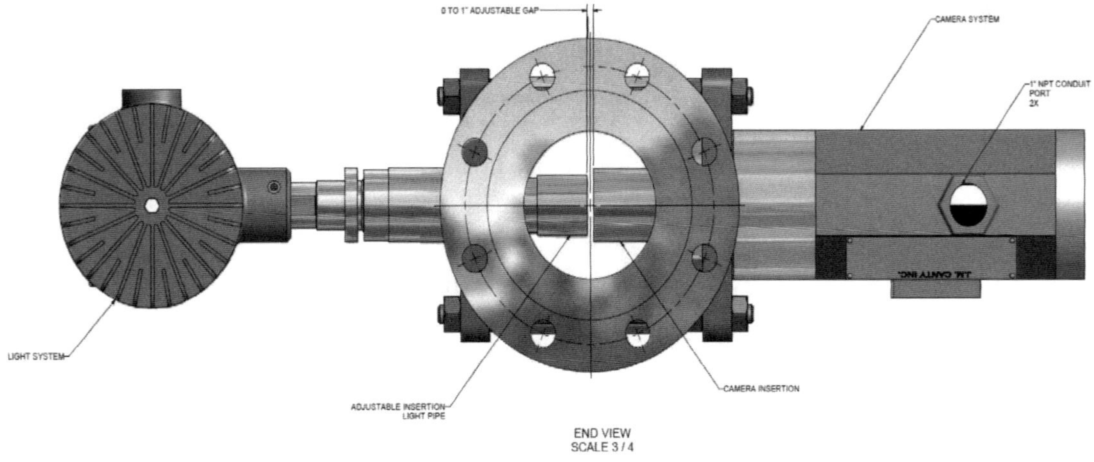

Fig. 7 - Inflow Systems ½" – 4" Pipeline

VISION TECHNOLOGY - SOFTWARE

The second point of discussion deals with the software. Vision systems are generally based on the same concept as the human eye – brain interaction. Objects are seen in our field of vision because they stand out from the background. This delineation is created by differences in light intensity around the object caused by shadows or colors. The vision system detects particles in much the same way. Intensity differences from the background field occur at particle boundaries. The vision system detects these and is able to form the particle based on defined threshold information which sets the level of intensity change that defines a particle boundary. The threshold information is input by the operator based on what he sees on screen. This is the powerful advantage of vision over other types of instrumentation. Reliance on indirect calibration is not required. The actual particles are used to define what is a proper size detection and what is not. This is depicted in the Fig. 8-9.

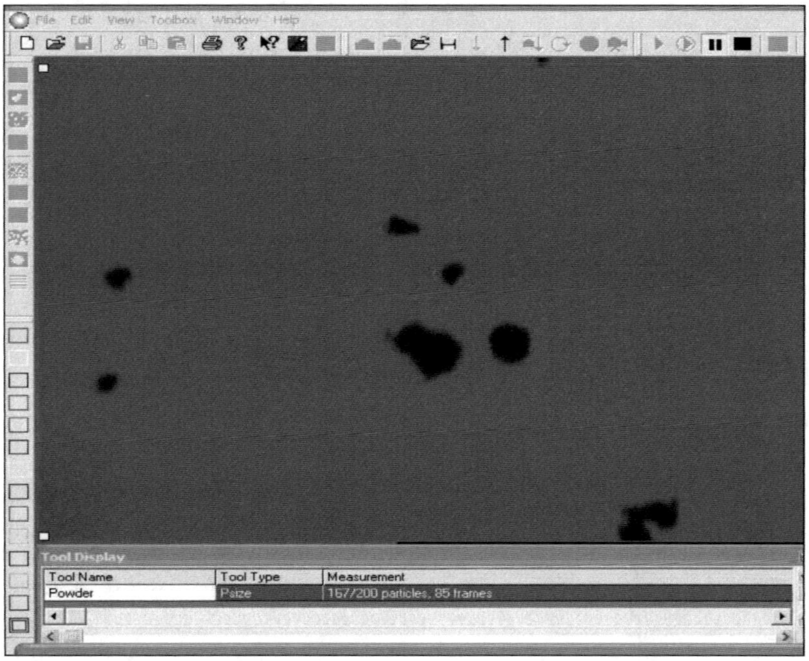

Figure 8. Live Stream View

Figure 8 shows an image of oil particles in the stream which can be used to define to the software what the particles actually are. The next step is to digitize the image and test thresholds to properly define what a particle is based on the actual view of the particle. Figure 9 shows a digitized image of the particle capture which yields particles as they are actually seen. This ability to select the threshold is a way the operator can confirm that the software is capturing particles correctly for analysis.

Fig. 9 – Proper Threshold Selected

The discussion to this point will serve as a basis for the remainder of this paper. The hardware and software principles are largely the same regardless of applications. The discussion to follow will focus on the data that can be extracted from the process image to enhance the production and control of the aforementioned applications.

DETECTION OF WAX and HYDRATES

The presence of precipitated wax in a flow line or pipeline can lead to clogging issues. Detection of wax precipitation can alert the operator to the pending problem and prevent a stoppage in production at best, and a loss of a flow line at worst. As process temperature cools, the wax takes on a solid form which tends to adhere to metal surfaces. Build up over time will choke the line. Figure 10 shows results of the system's ability to track wax precipitation in field trials as a function of temperature. Figure 11 shows a typical image of a crude oil which contains water and precipitated wax. The circular ring shapes are water droplets and the semi transparent, large formations are wax particles.

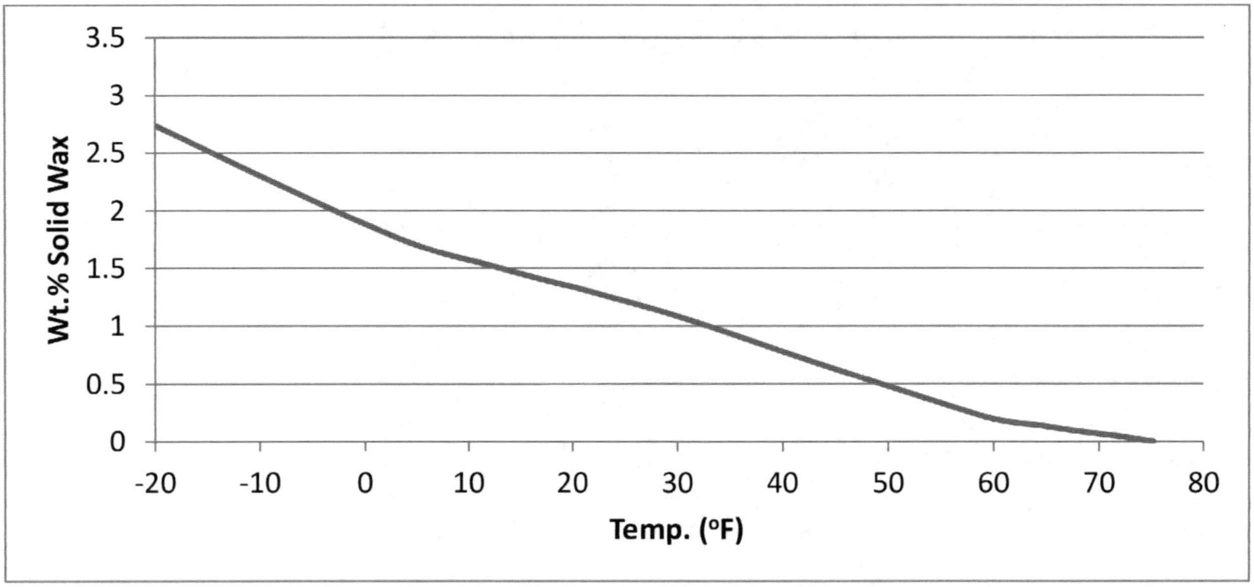

Figure 10. Wax vs Temperature

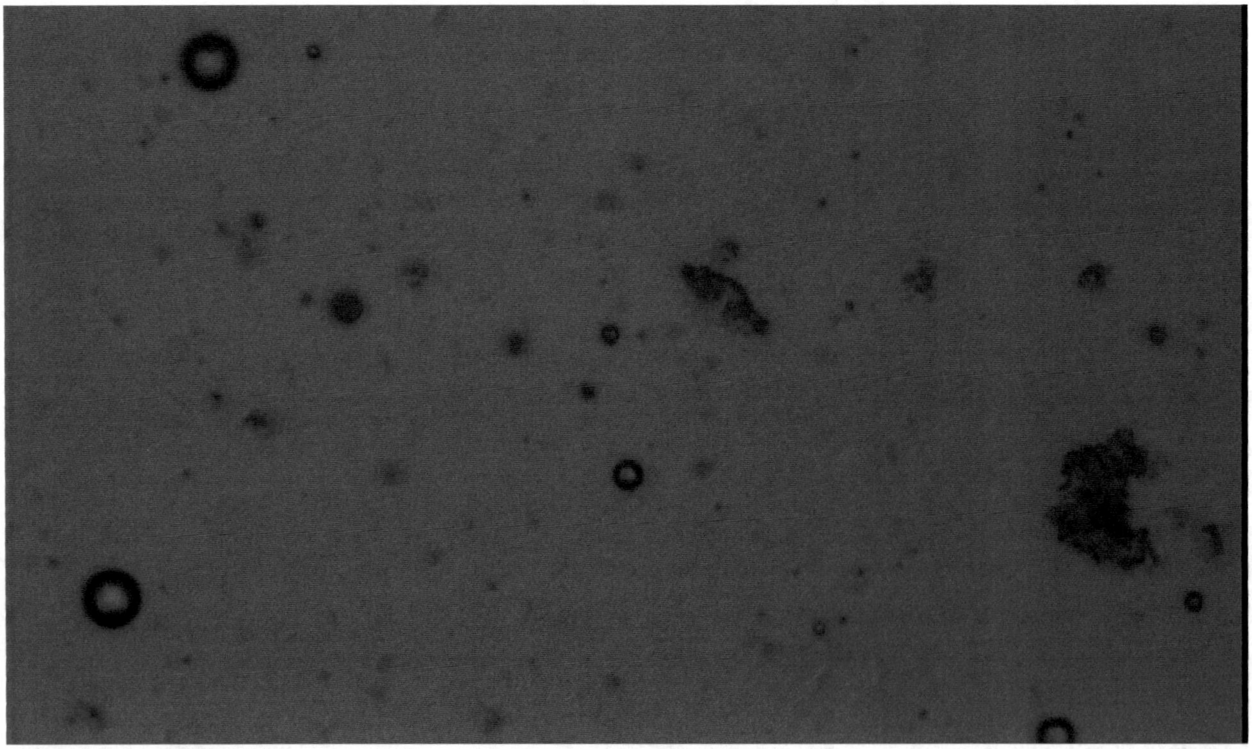

Figure 11. Typical Image of Crude with Water and Wax

Hydrates are detected in a similar fashion as they crystallize out of solution. Methane droplets will start to freeze, then agglomerate and finally plug the line if left untreated long enough by either temperature control or chemical injection. As their formation proceeds they will obtain an opaque shell which is indicative that the process is well underway and action must be taken to avoid disaster.

DETECTION OF SAND

The detection of sand is a critical function in the separation process. Once removed from the oil it must be cleaned, injected or dumped if present in small enough quantities. Detection is critical in any case. Viewing a produced water image one will see oil droplets, sand and other contaminants. General acceptance levels to dump to the environment are 30 ppm which is not a lot. When sand is detected in the outflow of the separator it is an indicator that separation must be adjusted to handle the present characteristics of the crude. Unlike oil droplets, sand, and other solids will almost always be non circular in the view. Vision software allows the categorization of these shapes for identification. In the image following a sand particle is seen in the center and upper left corner. These can be identified by aspect ratio or circularity as an example. Concentration of each constituent can be determined.

Figure 12. Sand Particles in Produced Water

CONCLUSION

Vision technology has the ability to maintain its viability in the harsh oil environment through design and manufacturing inherent in its components. Good visual optics, lighting and control of the measurement zone are all critical elements to success. Software has the advantage over other technologies due to its capability to analyze shape as well as size. This makes the difference between counting particles and classifying particles for a full understanding of how the process is behaving.

Overview of P.E. Licensing Process in United States

William R. Parrish, P.E. and Cory D. Jensen, P.E.

Professional engineering licensure was established in the United States to protect the health, safety and welfare of the public from unscrupulous or incompetent decisions of engineers. Licensure provides a proven and accepted assessment of experience, education and ethical standards.

This presentation gives an overview of the benefits of being licensed, the key players involved in becoming licensed and the path to licensure. The overview presents the background information to help put the other session presentations in perspective. More details about licensure are available in the March issue of *Chemical Engineering Progress (CEP).*

Value of being licensed

Being licensed shows other that you have a strong commitment to high technical and ethical standards. It can provide you with better career opportunities and provide a competitive edge when seeking employment. Many employers state that they would choose a licensed engineer over an unlicensed engineer, given both are otherwise equally qualified. Also, companies often offer incentives to encourage their employees to become licensed.

Some engineers think that licensure is important only in consulting or expert testimony. However, licensure is often required in firms doing design, construction and startup of processes. State laws also may include those teaching "advanced" engineering subjects be licensed.

Some states allow an "industrial exemption" for large companies who have the financial resources to cover liabilities arising from employee decisions. Most companies encourage licensure but many oppose eliminating the industrial exemption if they have operations in multiple states because it requires their engineers to be licensed in multiple states. There is a push to eliminate the industrial exemption. This important topic is discussed in another presentation.

No one can predict the exact future course of their career. However, being licensed may provide unexpected rewards and opportunities.

Key players in licensure

Becoming a P.E. involves education, experience and examination. Besides your coworkers, who can help guide your job experience, there are two important players in becoming licensed. The first is your local engineering board that establishes the standards for education and experience and ability to take the examinations. You will work with the board throughout the licensure process.

The second important player is the National Council of Examiners for Engineering and Surveying (NCEES) (www.ncees.org) . This is a nonprofit organization whose members are the various engineering licensing boards. The NCEES was created to provide some uniformity of standards of education, experience and examination. The membership establishes Model Law to offer uniformity. However, Model Law is a set of suggested best practices and no board follows Model Law exactly. All exams are administered and graded by the NCEES.

The licensure process

Obtaining a P.E. involves the following steps:

1. Having the appropriate formal education
2. Passing the Fundamentals of Engineering (FE) exam
3. Obtaining the appropriate experience (engineering internship)
4. Providing required documentation to board
5. Passing the Chemical Professional Engineering (PE) exam

While the general requirements are uniform across all jurisdictions, exact details vary. It is critical to work with your local board so that you understand and meet the exact requirements. A brief synopsis of each step is given below. More details are provided in the *CEP* article and in other presentations in this session.

1. Formal education: All jurisdictions require BS engineering degree, or equivalent, from an EAC/ABET-accredited four-year engineering program. Foreign degrees and other degrees are considered but must be approved by the board.
   Current Model Law states that as of 2020, engineers taking the PE exam must have a Masters or equivalent (MOE). This requirement is strongly opposed by AIChE and many other organizations. More details are presented in another presentation in this session and in the *CEP* article.

2. FE Exam: This exam tests your fundamental knowledge of engineering. It is easiest to pass when you are first eligible to take it (usually senior year or shortly after graduation). Another presentation and the *CEP* article provide more details of the exam and the upcoming changes in the exam.

3. Practical experience: Your work experience must be appropriate engineering practice. Typically four years are required although some boards permit fewer. In some jurisdictions advanced degrees can reduce the number of years required.

4. Documentation: The exact format of documentation and what is required varies between jurisdictions so it critical to work with your local board on this. However, all will require proof of education, passing the FE exam and work experience. The work experience will be supported by several references, who will also vouch for your ethical behavior.

5. PE Exam: This is the final step in becoming a licensed professional engineer. The exam is designed to be passed by a "minimally competent engineer with four years of practical experience". This exam tests the practical application of your engineering knowledge. Both a session presentation and the *CEP* article provide more details on the exam.

Summary

The AIChE recommends that its members become licensed to help ensure a competent and ethical chemical engineering workforce. While the process may seem daunting at first, working with your fellow employees and your local board will make the effort much easier. Each year about 400 chemical engineers become P.E's. They have made a significant investment in themselves and possibly in their future career.

(78b) Licensure--An Academic Perspective

S. Ranil Wickramasinghe, Department of Chemical Engineering, University of Arkansas, Fayetteville, AR

Most Chemical Engineering faculty members are not licensed. Why? Is it important for faculty to be licensed? Most undergraduates take the Fundamentals of Engineering exam but do not take the Professional Engineering exam. Why? Should faculty be licensed? Is licensure important for faculty who are conducting research? Should we encourage faculty to obtain their PE license? In this presentation these questions will be discussed.

TESTING FOR COMPETENCY - THE PROFESSIONAL ENGINEERING EXAMINATION

Anne Bertelsmann, PhD, PE, Galveston Bay Refinery, Marathon Petroleum Company LP, Texas City, TX
And
Denise Chastain-Knight, PE, CFSE, CH2M HILL, Augusta, GA

Introduction

The decision to pursue engineering licensure is a personal one, and the road includes many hurdles to overcome. Engineers take this road for a variety of reasons, with different motivations and experiences along the way, and finish the trek with the PE Exam. The Principals and Practice of Engineering Examination (PE Exam) is a nationwide exam used to ensure that licensed PE's are "minimally competent". The exam must include test content applicable to the competency being measured, effectively differentiate between competent and not competent candidates, and provide a consistent competency measure from administration to administration. This paper will define minimal competency, discuss the purpose of the Professional Engineering Exam, exam format using the Chemical Engineering Exam as the example, development methodology, item quality metrics and offer suggestions for test taking preparation. It will also describe the function of the NCEES Exam Development Committee and their role in developing a fair, effective exam, and explain why it takes weeks to get a result for a computer scored exam.

Path to Licensure

Whether an engineer undertakes the path to licensure early or late in their career, each must successfully negotiate the same hurdles. Each step in the process has a specific goal to either develop or to measure the engineer's capabilities toward professional competency. Figure 1 on the next page illustrates the path to licensure.

The quest to licensure begins with obtaining a Chemical Engineering degree. The next step is to pass an exam that measures competency on engineering basics. This first exam is the Fundamentals of Engineering Exam (FE) – also sometimes called the EIT (Engineer in Training) Exam. The FE exam is typically taken either during the senior year in college or shortly after graduation. The types of items resemble "textbook" questions and are designed to test understanding of the engineering fundamentals as taught in a typical engineering degree program. Currently, the FE examination consists of a general morning session (common to all disciplines) and a choice of seven discipline specific afternoon modules. Starting in the beginning of 2014, the FE examination will convert to a computer based test, which is discipline specific (i.e. no general module). For more information on the FE exam content and the conversion to computer based testing, refer to the NCEES website (http://ncees.org).

Figure 1: Path to PE Licensure

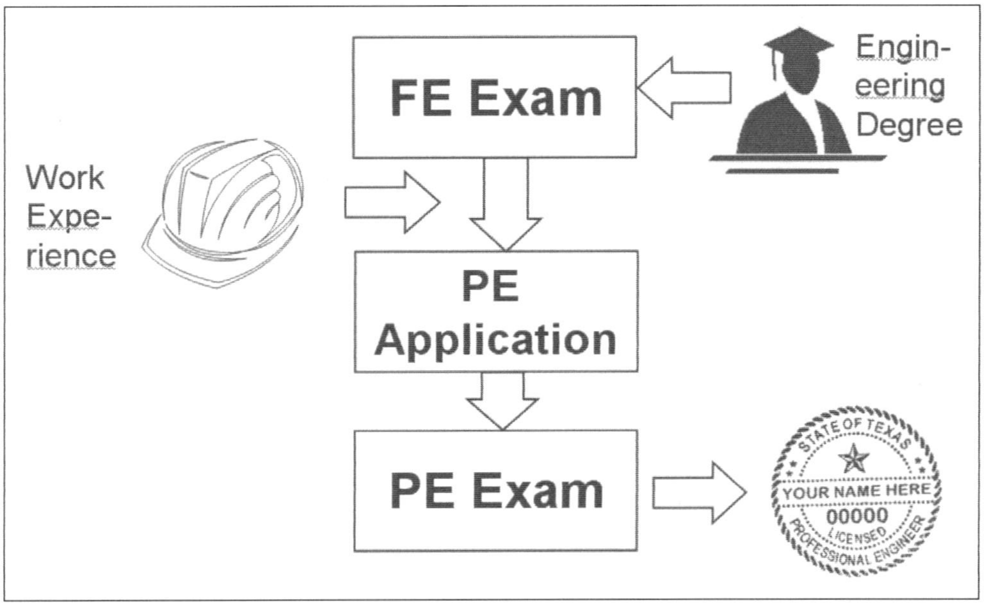

Following graduation, engineers enter the practice phase of their careers where fundamental knowledge is applied to real world application. In this phase of their careers, engineers use the theoretical principals learned from their education to solve problems that ultimately deliver physical products to market. Most states require a minimum of 4 years engineering related work experience before allowing candidates to sit for the PE exam. In order to become licensed, the candidate needs to submit an application to the state licensing board, describing education and experience, documenting passing of FE exam, and providing character and skills performance references. Format of the application and requirements for references, transcripts etc. vary from state to state. Once the application is accepted by the board, the candidate can register for the PE exam. After passing the PE examination, the engineer is licensed and can obtain the PE stamp. It is through this process that the state licensing boards ensure that only qualified candidates become licensed.

PE Exam Purpose

The PE exam is the final step in the road to licensure and its purpose is to ensure that only competent engineers become licensed. This helps to uphold the high standards of licensure and protects the health and welfare of the public. Since the exam is the same nationwide, it ensures consistency between states. Thus, one of the main requirements for a PE examination is that it can distinguish between competence and incompetence and that it can demonstrate this reliably and consistently. It is important to remember that "competent" should not be equated with "expert". The purpose of licensure is to identify individuals who have at least the minimum knowledge at which a person can practice professional engineering in such a manner that will safeguard life, health and property and promote the public welfare. That is why the term "minimally competent" is used.

A minimally competent engineer:

- should be able to perform identified tasks
- should meet minimum education and experience requirements
- is beginning to specialize in a sub-discipline
- may safely practice in another sub-discipline even though he/she does not practice in this area on a daily basis

The definition of "minimally competent" has implications on how the questions on the exam are written. The PE exam differs from the FE exam in that it addresses both the principles and the practice of engineering. So, while the FE questions are typically textbook related, questions on the PE exam are predominantly practice related. The questions build upon the experience that an engineer has gained during his or her first few years of working and need to recognize that chemical engineering experience can be gained in a variety of areas. The difficulty level of the questions is geared towards an engineer with about four years of experience.

PE Exam Content

The content of the examination is determined through a survey of practicing chemical engineers (PAKS study = Professional Activities, Knowledge and Skills), which is repeated about every 6-8 years to ensure that the examination stays current. The PAKS process includes conducting a technical content survey with input from practicing professionals. This survey is conducted over a period of weeks to ensure that the respondent pool size is adequate and is sufficiently diverse to be representative of the overall profession. Survey results are analyzed by a team of licensed Chemical Engineers, most with no prior exposure to PE exam development work, under the direction of a psychometrician who ensures that proper procedures are followed. The PAKS team will determine both the domain (subject area) content and item weight of the subdomains. PE exam content is established based on the PAKS study, and is published in the format of the PE exam specification available from the NCEES website (http://ncees.org/exams/pe-exam/) and is summarized in Table 1 on the next page.

The ChE exam specification was recently updated and became effective for the April 2011 exam administration. It includes six major domains and fifteen sub-domains that investigate theoretical and practical knowledge. Exam content is distributed within the major and subdomains according to the percentages shown in Table 1 to ensure a consistent content basis for every exam. The specification also includes guidelines for topical content within each sub-domain. For example the exam will include 4% of the items from sub-domain VI. C. Operation. Figure 2 includes example content from the 2011 exam specification.

Table 1: April 2011 ChE PE Exam Specification Summary

Knowledge Area	Approximate Percentage of Examination
I. Mass/Energy Balances and Thermodynamics	**23%**
A. Mass Balances	10%
B. Energy Balances and Thermodynamics	13%
II. Heat Transfer	**16%**
A. Mechanisms	9%
B. Applications	7%
III. Kinetics	**11%**
A. Reaction Parameters	6%
B. Reactors	5%
IV. Fluids	**16%**
A. Mechanical-Energy Balance	12%
B. Flow and Pressure Measurement Techniques	4%
V. Mass Transfer	**14%**
A. Phase Equilibria	5%
B. Continuous Vapor-Liquid Contactors	8%
C. Miscellaneous Mass Transfer Processes	1%
VI. Plant Design and Operation	**20%**
A. Economic Considerations	1%
B. Design	10%
C. Operation	4%
D. Safety, Health, and Environment	5%

Figure 2: Sample Exam Specification Sub-domain Content

C. Operation 4%

1. Process and equipment reliability (e.g., testing, preventive maintenance, startup/shutdown procedures, robustness)
2. Process improvement and troubleshooting (e.g., debottlenecking, experimental design and evaluation, optimization)

PE Exam Format:

The PE examination is given twice a year - once in April and once in October. It is an eight hour exam consisting of a four hour morning session and a four hour afternoon session with a lunch break in between. There are a total of 80 multiple choice questions (40 in the morning, 40 in the afternoon). Each question has four answer choices of which one and only one is the correct answer. Each question is stand-alone, i.e. it is independent of the other questions on the exam and can be solved with the information given in the problem, sometimes with the help of standard references. Some questions involve calculations while others are qualitative. All questions count equally and there is no penalty for guessing.

Each of the 80 multiple choice questions on the exam is of the same general format. In exam development, questions are referred to as "items" rather than questions because not every item has a question mark (like the one in the example below). Figure 3 illustrates the anatomy of an item, i.e. its general format and identifies the different parts of the item.

Figure 3: Anatomy of an Item

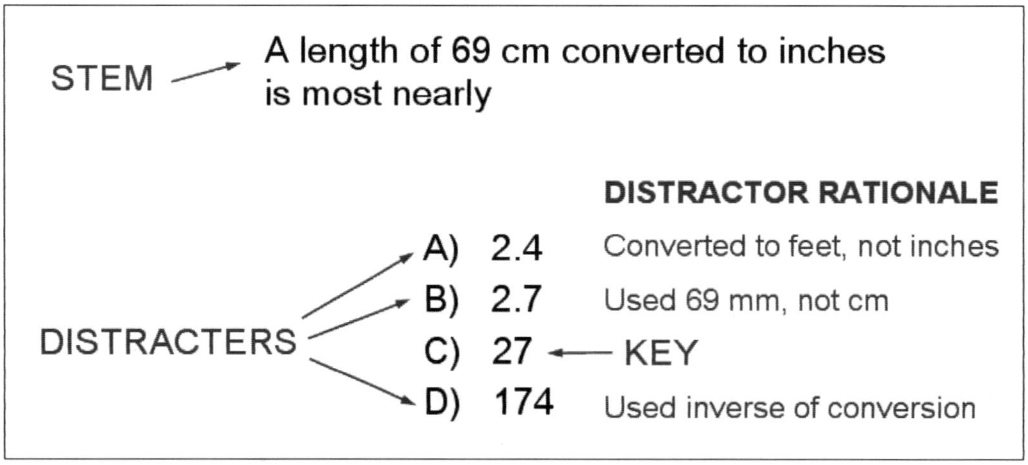

Each item consists of a stem and four answer choices. The stem (or problem statement) identifies what calculations are needed. It includes the information that is provided to solve the problem (sometimes even in table format) and states what is required for an answer. Beneath the stem are the four answer choices. Of these four answer choices, one (and only one) is the correct answer – the key. The other three are so-called distracters. They are incorrect answers – but they are not random numbers.

Each of the wrong choices represents a typical error that a less than minimally competent engineer (or someone not familiar with this specific topic) may make. Each distracter has to have a "rationale", i.e. an explanation which error was made. So, when a candidate is taking the exam and calculating a number that is exactly one of the given answer choices it does not necessarily mean that the problem was solved correctly. Maybe the candidate just made one of the "expected likely errors"…

The distracters can have a significant impact on the statistics of an item. If one of the distracters is obviously wrong (such as a negative absolute pressure), then even less than competent candidates will disregard it. That reduces the feasible choices, improves the "guessing" performance, and increases the chance that a not minimally competent candidate gets the question right. The Exam Development Committee strives to develop items that include the key, and three equally likely errors as distracters.

Exam Development

The PE Examination (as well as the FE, and the Surveying Exams) is a service that NCEES provides to the licensing boards. As such, NCEES provides the framework and the oversight for the exams, but the responsibility for the content rests with each exam committee. The committee writes the items, assembles the exams and reviews the statistics. NCEES has established rules and procedures that the exam committees follow to ensure consistency. Committees report back to NCEES with their exam quality measures. NCEES also provides the logistical support (psychometric consultant, publishing and printing, administrative support, maintaining records and the item bank, administering the exam, scoring, etc) and provides all security functions. For the larger exams, NCEES also pays and arranges for volunteer travel. For the smaller exams, this is the responsibility of the sponsoring society.

The Chemical Exam Committee is the group that actually writes and reviews the Chemical PE Exam. To be on the committee, one must be a licensed PE and express interest through the NCEES website or through a committee member. The committee has about 50 active members and meets twice a year in Clemson, South Carolina at the NCEES headquarters. Not everyone comes to every meeting, of course, so typically 30 – 35 committee members present at any given meeting. During the meeting items are written and reviewed by committee members who work with one another to critique and improve the items. Task teams also work to assemble, review and pretest exams.

Before an exam is administered, each item will have no less than six documented formal reviews and likely have been assessed by more than 20 people. Committee members are asked to complete detail oriented and often challenging engineering work and can earn PDH (professional development hours) for this activity. Developing the PE exam helps committee members to keep their engineering skills current and to give back to the profession. Without the work of this committee, the PE exam and with it the licensing wouldn't happen.

Preparing an exam that effectively and consistently measures minimal competency is a challenging task that succeeds or fails based on the ability of the committee to predict item performance. Each committee relies on a structured process and employees a variety of quality tools and techniques to improve the opportunity for success. Figure 4 is a workflow of the exam development activities. Items are written by a PE and submitted for review. Many items are prepared by the committee members at each meeting, while others are submitted by PEs who do not attend the meetings. The individuals who write and review items are referred to as Subject Matter Experts (SME). The author and two reviewers will independently work the item and develop the rationale for each distracter. Through this process the item is often revised and edited to improve its presentation and quality. This step is considered one of the most critical activities of the committee because items are the building blocks of the exam and each must meet high quality standards. Each committee member is responsible for item quality and is granted authority to modify draft items as needed to meet quality objectives, or reject items that fail criteria and cannot be edited to usable items. Finally the item is given a final check by the committee chair (or authorized designee) to verify that documentation is complete, both reviewers agree to the item content, and all quality checks have been completed, and approves the item for acceptance into the item bank. The item bank is a repository for items that are ready to be placed on exams.

Figure 4: Exam Development Process

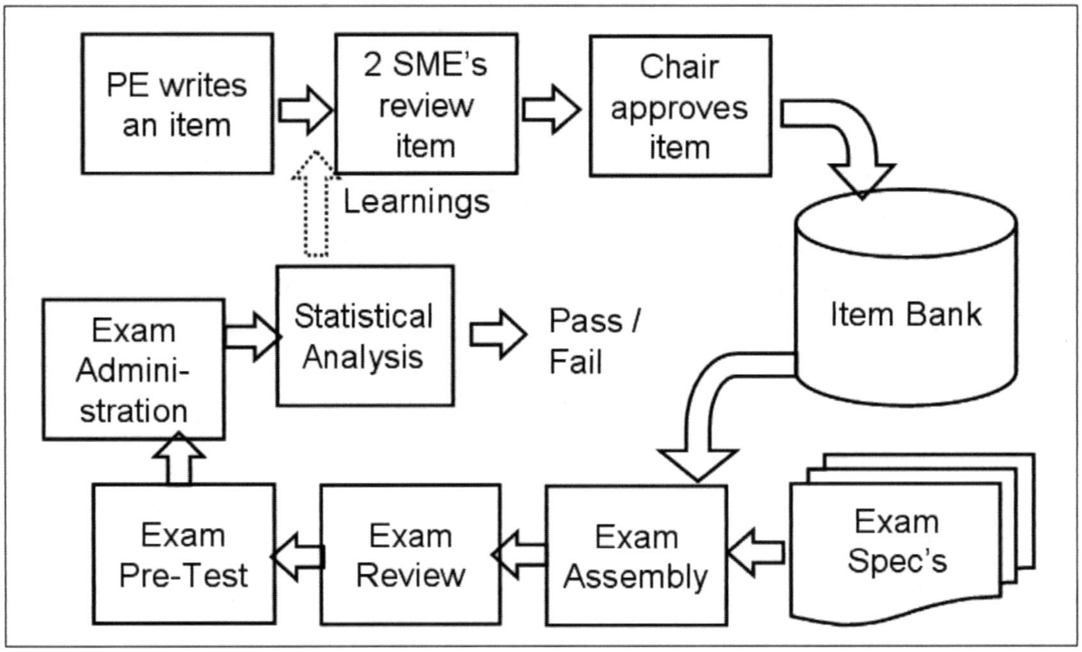

Exams are built, reviewed, revised, reviewed, quality tested and reviewed over a series of meetings. Each exam will have three independent team reviews, a pretest and a quality assessment. There are specific checks required at each workflow gate with task teams assigned to maximize the number of eyes that evaluate any given exam. An exam is compiled using items from the bank following the content distribution of the specification and sequenced in the specification order. The build team will then conduct a complete review of the exam, looking at each exam item for individual quality and checking for overall content flow and replacing items if necessary. The exam is then reviewed in two more meetings by independent teams who provide a "fresh set of eyes" and continue to improve the content to ensure that each item on the exam and the overall exam meet quality criteria. They also verify that the documentation is complete in accordance with the NCEES exam development requirements. Each exam must also have a pre-test where a group of committee members take the exam in as if they were actual test takers. After working all the items on the exam, they answer a series of questions about the items designed to predict item performance, then perform a distracter evaluation. Comments from the pre-testers are incorporated into the exam. In the final meeting before administration, a quality team conducts one last review of the exam verifying that all preceding checks (and associated documentation) are complete, all comments have been addressed, specification distribution is correct, and that author diversity limits and other NCEES criteria are met.

When the exam is administered and scored, data is compiled that can be used to evaluate the quality of the exam as a whole, and individual items. The data is analyzed and used to help the committee to learn what makes items perform effectively, and improve our performance prediction skills. Authors are provided a 'report card' so they know how items they wrote performed and have the opportunity to see how those items evolved in the review process. At each meeting, the committee will review items that had great statistics and try to find out what made them perform well. Similarly, we will review items that didn't perform well and identify traits that are undesirable in future items.

Item Statistics

Item quality is ultimately measured by the statistics generated from the candidate responses after an exam administration. Before discussing the statistical analysis of the examinee responses, one must first consider what makes an item a "good" item. The definition of a good item is one that is capable of distinguishing between "minimally competent" and not "minimally competent" candidates, does not discriminate against any particular group of test takers, and performs consistently in multiple exam administrations. To achieve those goals, it must be relevant, not confusing, and measure competency.

Most importantly, the content of the item must be relevant. Relevance is defined as the appropriateness of the topic and the difficulty of the item. The topic of the item must test knowledge or skills that are directly related to the exam specifications. For example: a question on high voltage power distribution may be relevant for an electrical engineer, but will not provide information on engineering competency of a *chemical* engineering candidate – no matter how well written it is. If the item does not relate to the exam specifications, it does not belong on the exam. Second, the difficulty level and complexity of the item must be such that a "minimally competent" engineer with 4 years experience can solve it in about 6 minutes. When writing items, authors often think about the interesting and challenging problems that are encountered in practice; however, if an engineer with 20 years experience requires several hours or days to solve the problem, what are the chances that a junior engineer will get the answer in 6 minutes? These type of problems, as interesting as they may be, do not make good exam items.

Another important principle to consider is that items should not be confusing to people who know the answer. When an item is confusing, a competent engineer who understands the concept being tested may not select the key, thus the question does not help to distinguish between competent and not competent. The so called "trick" question does not evaluate a candidate's competence and is considered a "bad" item to be weeded out in the exam development process.

It is equally undesirable to use items that give the answer away to people who don't know the answer. If the item includes clues to the key, then an incompetent engineer can choose the right answer without understanding the subject matter; thus the item does not measure competency. Clues can take many different forms:

- If some of the answer choices can be ruled out as impossible (like negative absolute pressures), the chance of guessing the right answer goes up.
- Grammar can give clues when one of the answer choices "stands out", i.e. is more detailed or does / does not match the grammatical structure of the item statement

For example, an item that asks about the 19th president of the United States, where three of the four answer choices are well known pop stars, does not require the candidate to know anything about US history to pick the correct answer.

Following an exam administration, a psychometric analysis is performed (psycho – metric = measuring the mind) on examinee responses. The analysis is solely based on information provided on the marked answer sheets without regard to the actual content of each item, but quite a bit of information about an item can be observed just from examinee responses. The psychometric analysis will evaluate a number of things. Of course, the simplest analysis is to determine how many candidates selected the key. Statistically, there is a 25% chance of guessing the correct answer because there are four choices. An item where less than 25% of the candidates selected the key may be flawed because it may have directed candidates towards preferentially selecting a distracter over the key. The next question is to ask *who* got the question right. Ideally, the candidates who got the question right were the same people who performed well on the exam overall. The analysis looks at the distribution of answers between the distracters and who selected them. This analysis also considers if all distracters were selected equally or if there was one distracter that was particularly attractive; maybe it even got selected more frequently than the key.

The objective of this analysis is to confirm the validity of each item as a tool capable of measuring engineering competence. An item that follows the basic three criteria (relevance, no confusion and no clues) can be expected to distinguish between competence and incompetence; however, sometimes clues, confusion or complexity can creep into an item unintentionally and be missed during the review before the exam is given. Statistics are then used to identify potential flaws. One such flaw is a mis-key. Mis-key is when the committee identified for example option A as the correct right answer, but option B is actually correct. Another flaw is an ambiguous item where more than one of the options could be correct. Despite the committee's efforts, such flawed items occasionally make it through the review process and are placed on an exam. The rigor of review process prior to exam administration minimizes the likelihood of a flawed item; however, it cannot totally eliminate them. The statistics after the exam will reflect item imperfections and provide an opportunity to compensate for item flaws in exam scoring so that candidates are not negatively impacted by flawed items.

Figure 5 is an example of the statistical analysis report for an item prepared by the psychometric consultant. One such sheet is prepared for each of the 80 items on an exam. The top left corner shows the "number of cases", i.e. the number of examinees who took the exam in this administration. In this example, there were 2910 candidates. So, this example is not from any Chemical PE exam where typically about 300 candidates sit per exam. The number below the case number (highlighted in yellow) is the Mean (or the P+ value). This number shows what percentage of the candidates selected the key; in this case, 65%. P+ alone indicates that the question is of appropriate difficulty for the average test taker; not too easy and not too hard. Statistically, the higher the P+, the easier the item.

Figure 5: Item Statistics

Number of Cases	2910
Mean (P+)	0.65
Delta	11.44
Pearson	0.33
Biserial	0.43
Mean of Passers	0.78
Mean of Failers	0.53
Mean Near Cut	0.68
Phi	0.26
DIF Grade	

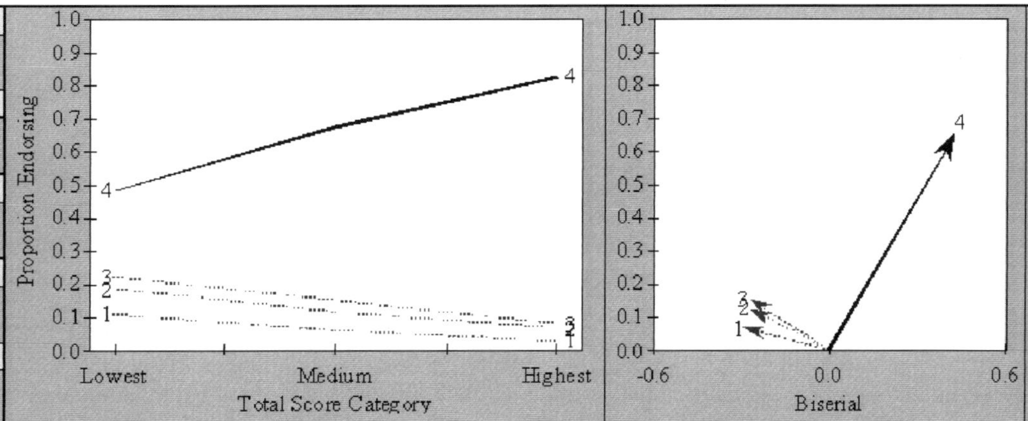

Alt.	Lowest	Low	Medium	High	Highest	%	N	Mean Scr	Biserial
1	0.11		0.06		0.03	6.7%	195	43.0	-0.29
2	0.19		0.12		0.07	12.5%	365	44.1	-0.27
3	0.22		0.15		0.08	15.4%	449	44.3	-0.27
4*	0.48		0.67		0.83	65.2%	1896	50.1	0.43
OM	0.00		0.00		0.00	0.2%	5	42.4	-0.20
Mn Scr	38.68		48.50		58.27				
Grp N	1025		1022		863				

The statistics not only evaluate how many candidates selected the key, but how selecting the key on this item correlates to overall exam scores. Test takers are divided into three groups of roughly equal size based on overall score on the exam; the low scoring third, the medium scoring third and the high scoring third. For each of these groups, the report in the table under the left graph shows how many selected which answer choice. For example on this question, “4” (or “D”) was the key as indicated by the asterisk. Of the low scoring third, only 48% got the right answer, while 83% of the high performers selected the key. Thus, someone who performs well on the exam overall (which is an indication of competence) is significantly more likely to select the key than someone who scored low (may indicate lack of minimal competence). This spread shows that the question is capable of testing competency.

Figure 5 also illustrates that the incorrect answers were relatively evenly distributed between the distracters, indicating that none of the distracters appears to be overly attractive or easily ruled out. Another way of looking at the same data is presented on the right of the page. The table shows average overall scores for the people who selected each of the answer choices. In this case, the people who selected the key had an average overall score of 50, while the people who selected the distracters had overall scores in the range of 43 to 44. Five candidates with an average score of 42 did not answer the item. This clearly demonstrates that the people who selected the key were more likely to do well on the exam overall, and indicates that the item can effectively measure competency.

The data is evaluated in yet another way by the cumulative statistics in the lower left: Mean of Passers indicates that of those candidates who passed the exam, 78% selected the key. Mean of Failers shows that only 53% of the people who failed the exam selected the key. The larger the spread between these two numbers, the better the item is at measuring competency.

A handy way of summarizing the "who got the question right" analysis is the R-Biserial number. Basically, this number is related to the slope of the lines in the graph on the left. For a good item, this slope for the key should be positive and the larger the better because it expresses the difference in performance of the low and the high scorers. Theoretical range for the R-Biserial number is from -1 to +1, positive values indicate that the candidates who selected the key also scored higher on the overall exam. A negative R-Biserial number indicates that higher scoring candidates selected one of the distracters, which could be an indication of a mis-key or ambiguous item. The R-Biserial number is highlighted in yellow on the left and shown in the picture on the right. The R-Biserial number is important because it evaluates how well an item meets the three basic criteria. We previously observed that the sample item is of appropriate difficulty based on the percentage of correct answers (P+ value). The R-Biserial number shows that we did not confuse the competent candidates (a high percentage of high performers got it right) and apparently it did not give any clues to the less competent candidates (. It also appears to be relevant with regard to the topic since the high performers (those with better understanding of chemical engineering) did better on this than the low performers.

To further illustrate the concept of the R-Biserial number, consider an example of an item that might produce undesirable statistics. The item is asked on the final exam for a graduating class of chemical engineers: "Which bar in town has the most beers on tap?" One would expect a certain percentage of students to select the correct answer, but are those candidates the most knowledgeable students who spend a lot of time in the lab, or the "party crowd"? This question would likely have a negative R-Biserial number, indicating the overall least knowledgeable students are more likely to select the key, which means this item is not an effective measure of a graduating chemical engineer competency, nor is it relevant to chemical engineering knowledge and skills.

Determining Competency

The individual item statistics are the building blocks for determining the overall exam performance. To make the move from the individual item statistics to determining competency of the candidate, a passing score, also called cut score, needs to be established. Figure 6 on the next page illustrates the philosophy behind relating the score distribution to the cut point, or passing score. Total score is simply the number of items answered correctly. For the Chemical Engineering PE Exam, the total score has a theoretical range of 0 to 80. Most candidates have at least some knowledge and get lucky guessing on some questions, so scores on the very low end are uncommon.

If it were known upfront which candidates are competent and which ones are not, the scores of the two groups could be plotted as shown in Figure 6. The at least minimally competent engineers have higher scores, the not minimally competent engineers have lower scores. However, there will be some overlap between the two groups. The better individual items differentiate between competency and incompetency, the smaller the overlap. However, because there is a range of competency and because a person's performance can vary, there will always be some overlap and thus the cut point decision is not clear cut.

Figure 6: Score Distribution

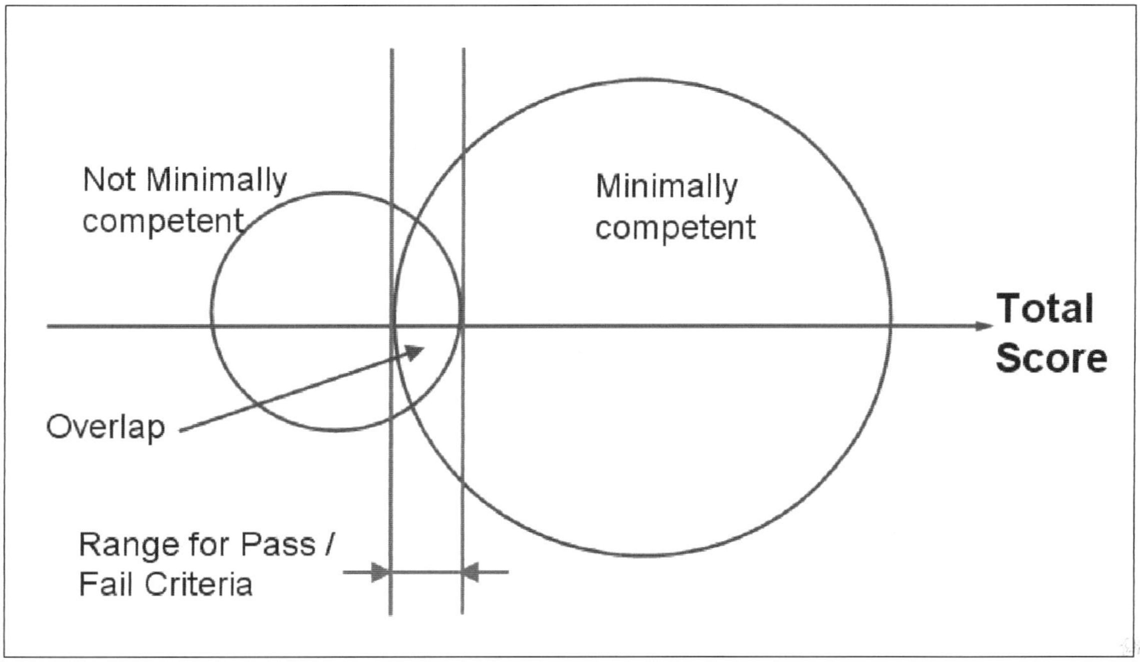

The overlap represents the range where the pass/fail decision should be made. All the at least minimally competent candidates on the right side should pass, all the not minimally competent candidates on the left should fail. With any decision point within that overlap range, there is a potential for competent engineers to fail and/or for less than minimally competent engineers to pass. Failing a competent engineer means that they will need to re-take the exam and includes the risk that they may give up on licensure. Passing a less than minimally competent engineer may allow an under-qualified person into public practice. Since the second type of error is more consequential, there is a tendency towards raising the bar when determining the pass/fail criteria.

Without knowing upfront which candidates are "minimally competent' and which ones are not (after all, it is the purpose of this exam to establish this), statistical measure are used to determine the pass/fail criteria, i.e. establish the "cut score". The cut score is the number of correct answers that a candidate needs to have in order to be considered "minimally competent". One way of determining the cut score is through panel of licensed PE's. This process is fairly involved, since it requires about 15-20 PE's who can volunteer 2-3 days of their time and travel to attend to cut score session. The panel must have good diversity (age, gender, area of practice, geographic location, demographics, etc…) and the cut score study must be completed before the results of the exam can be reported back to the state boards and ultimately to the candidates who are eagerly waiting for the scores. For these reasons, a cut score panel is usually only done when there was a change in exam specifications where the exam results cannot be directly compared to recent prior exams. The first exam where the format or specification is changed is called the "anchor exam" and cut point is established by a formal study.

For a cut score study the diverse panel of volunteers (all licensed PE's) meet and spend about a day to work the exam in a timed environment just like the actual administration. They can use any references, cannot consult with one another and they don't have access to the exam solution. The next day, they receive their score, and review and discuss each item as a group. They look at the statistics similar to those described above, but lacking the pass/fail metrics, review the solution, the distracters and discuss the topic and difficulty of each item. On the basis of that discussion, the panel members make a judgment call on how many "minimally competent" engineers out of 100 would answer this question correctly. A psychometrician guides the session and collects these estimates from each panel member. This data is then used in statistical analysis, generating a proposed passing score. The proposed passing score can then be adjusted (for example, to err on the "safe side"). This then becomes the basis for the pass/fail decision for the anchor exam.

Such a cut score study is not practical for every exam at every administration. So as long as there is no change in the exam specification, it is possible to "equate" the current exam to a previous exam. The equating process, heavily supported by a psychometrician, compares examinee performance on the current exam with examinee performance on previous exams. From this comparison it can be determined if the current exam is harder or easier than the previous exams, and cut point adjusted accordingly. If the data indicates that the exam is easier, the cut score is adjusted upward and a candidate will need to answer more questions correctly in order to pass. Conversely, candidates are not penalized if they take a harder exam because the cut point is revised downward and fewer questions must be answered correctly to pass. Equating is NOT the same as grading on a curve. When grading on a curve, the passing percentage would be fixed and the cut score adjusted to achieve that percentage. With equating, the competency of a candidate is measured on personal performance, and not dependent on the individual candidate's position within the test taking population, nor on the difficulty of the exam.

The psychometricians calculate other measures from the data. Coefficient Alpha measures the internal consistency or statistical homogeneity of the data, thus basically expresses the reliability of the exam. A measure called "Subkoviak" provides an estimate of the consistency of the pass/fail decision or an estimate of the probability that an individual would receive the same pass/fail decision on multiple administrations of the exam. All these measures are regularly reviewed to ensure that the exams can consistently measure competency.

After an exam is administered and all the answer sheets are scored, the statistical item analysis is performed. Exam Committees need to review potentially flawed items in case the key needs to be adjusted and the exam rescored. Then the cut point is determined, either through a cut score study with a panel or through equating. Only then can the pass / fail decision be made. State boards are notified with the results, perform state specific pass/fail assessments and then notify candidates of the results. NCEES recommends that only pass/fail should be reported to the candidates, though some states provide the actual score (or a "scaled percentage" based on a fixed passing percentage) to candidates. Candidates who fail get a diagnostics report so that they know how they performed in the various subject areas. NCEES recommends to only report the pass/fail decision for two reasons: The cut score can vary from exam to exam, so a candidate may pass one year and another candidate taking the exam another year may fail with the same score. The other reason is that the exam is designed to make a pass/fail decision on minimum competency. In that regard, the PE exam is different from an exam like the SAT, which is designed to measure "aptitude" over a wide range of abilities. The PE exam is not designed to measure ability accurately at the far ends of the scale. The PE exam does not distinguish between the "outstandingly competent" and the merely "extraordinarily competent engineer". Similarly, it cannot distinguish between the "totally incompetent" and the "absolutely clueless". Thus, a candidate who received a score of 75 out of 80 is not necessarily a better engineer than a candidate with a score of 72 out of 80 – both are "minimally competent" and pass.

The skeptical reader asks; "Does this really work?" Can an exam such as this really measure competency as the statistical "number magic" appears to indicate? The answer is found in the historical exam pass rates. NCEES does not report an overall pass rate for the exam, but breaks it up into a pass rate for first time takers and for repeat takers. Table 2 includes pass rates for exams given under the current Chemical PE Specification. The most recent pass rates for other exams can be found on the NCEES website.

One specific item to note about the data in Table 2 is that there is a gap between the pass rate for first time takers and for repeat takers. The pass rate for first time takers is typically rather high; in the 70 – 80% range for the Chemical PE exam. This high pass rate is because the examinee pool is already a pre-selected group. In order to be admitted to the exam, a candidate had to pass the FE exam, had to prepare an application with documentation of engineering experience, and had to be approved by the state board, a process that weeds out some of the less qualified potential candidates. Also, especially in chemical engineering where there are many jobs that come under the industrial exemption and do not require licensure, the people who decide to pursue licensure are already a non-random subset of all engineering graduates. This contributes to the relatively high pass rates for first time takers. The portion of the candidate population that passes on a repeat attempt includes in candidates that are minimally competent but may have been inadequately prepare in their first exam sitting.

Table 2: Chemical PE Exam Pass Rates

Chemical PE Exam Pass Rates	April 2011	October 2011	April 2012	October 2012
1st Time Pass Rate	78%	75%	74%	71%
Repeat Pass Rate	31%	29%	31%	20%

How to Prepare for the Exam

All the fascinating information presented in this paper aside, chances are the reader really wants to know how to prepare for the exam. Preparation philosophy will differ for a young candidate near the 4 year experience mark from that for a veteran engineer. One theory is that any minimally competent engineer can pass the examination without any preparation if he or she had unlimited time - time to look up references, read up on concepts and topics outside one's familiar area of practice, find the right equations and conversion factors, etc. But there is a time limit, and 6 minutes per question is not a lot of time.

Exam preparation should focus on building speed and efficiency for solving items. Through study and preparation, a candidate should become familiar with reference content, placing tabs to help find frequently used tables or compile a personal reference of frequently used information. Common equations and frequently used conversion factors should be memorized. Candidates should reactivate the habit of using math: solving equations, working with variables and using a scientific calculator instead of a computer. Speaking of calculators, it is strongly recommended to prepare for the exam using one of the approved models. Precious time can be wasted searching for the logarithm button on an unfamiliar calculator. Approved calculator models can be found on the NCEES website.

Exam readiness is achieved through lots and lots of practice. Preparation classes are useful for refreshing material basics, but are no substitute for rehearsal. A candidate should prepare by solving as many sample problems as one can find. By solving many, many problems, one will automatically remember the frequently used equations and conversion factors, enhance math skills and remember where to find information. The process also develops test day confidence because the items encountered on the exam have been encountered in similar form during exam preparation. Candidates should conduct one or more mock exam day(s) to practice pacing and practice maintaining focus for 8 hours. NCEES publishes a sample exam for each specification that has been developed by the exam committees in the same manner as a "real" exam. One author of this paper frequently recommends to potential exam candidates that they practice with readily available materials, complete a timed trial with the NCEES exam, then continue practicing with more materials.

Finally each candidate should prepare for the test environment and stage to be at peak performance on test day. Exams are administered in finite locations within a state and often require candidates to travel. Plan a travel itinerary that will allow arrival at the test site early, rested and unstressed. Have a satisfying but not heavy breakfast and make a plan for a relaxing lunch. Consider the test room conditions, dress comfortably and in layers. Be aware of the materials that are allowed in the test hall, and prepare accordingly. Candidates will be turned away without a valid government photo ID. Cell phones and other forbidden materials may be confiscated or admission refused. The exam is worked in a printed booklet, which contains sufficient space for working the questions and pencils are provided. Candidates need to bring an approved calculator, unapproved calculators will be rejected. The exam is open book, so candidates also need to bring any references that they want to use during the exam. References must be bound (no loose paper), but can be marked with flags. Review the test day policies on the NCEES website and plan accordingly (http://ncees.org/exams/exam-day-policies/).

Closing Comments

Professional Engineer Licensure is a personal decision and requires one to undertake a dedicated and challenging journey. Candidate should embark on this journey knowing that the exam development process is designed to ensure each exam measures for minimal competency in a consistent manner, and items are statistically validated. Finally, the factors influencing exam score processing timelines are understood. The authors of this paper encourage engineers to pursue licensure. We speak for ourselves and our fellow PEs when we say that nobody ever regrets the resources and effort spent obtaining a PE, and we appreciate the industry approval and instant expert acceptance granted by those two little letters following our names. P.E. Professional Engineer.

(78d) The Current Hot Button Issues in Domestic Professional Licensing

Amos Holt, Southwest Research Institute, San Antonio, TX

Amos Holt, a past president of ASME and current VP for Environmental, Safety and Quality Systems will review the issues currently receiving the greatest attention by the PE community which includes all of the various state and territorial licensing boards, the National Council for Examiners of Engineers and Surveyors as well as the numerous professional society stakeholders and educational organizations and societies. The discussuion will touch on the Model Law and its currect sections on the Masters or Equivalent (MOE) requirements as well as Industrial Exemptions. The various positions of key players covered including the position of a consortium of professional societies called "Licensing that Works" or LTW will be summarized. This group includes both ASME and AIChE and represents about 240,000 engineers thtroughout the world. Amos will tell you what you will likely be hearing more about in the next few years and how it might affect both current PE licensees and prospective PE licensees.

The California Divide – And Why You Need To Engage Your Legislators

Diane Spencer, PE, Lawrence Livermore National Laboratory, CA
and Emmett Miller, PE, consultant and AIChE Fellow, CA

Abstract

California is crippled by a great divide that has nothing to do with tectonic plates. California's professional engineering statute gives civil engineers free reign, allows electrical and mechanical engineers to work in their disciplines so long as that does not conflict with civil engineering, and acknowledges by title only that other engineering disciplines exist.

This division presents problems for engineers whether they have a professional engineering license or not. For example, Spill Prevention Control and Countermeasure (SPCC) plans are mandated by federal regulations. Such plans are prepared by chemical engineers all over the country, but NOT in California. SPCC plans are part of the exclusive domain of civil engineers.

A fire protection engineer was once told he could not design a fire suppression system because that was the exclusive domain of mechanical engineers. A chemical engineer was once told he could not submit a Risk Management Plan (RMP) (required by Cal EPA) without a civil engineer stamp because even the hazard analysis on a facility that uses chemicals is the civil engineer's domain. These are just a few examples where engineers are not allowed to work on projects that similarly trained engineers are expected to do in the rest of the country and the rest of the world.

So how do engineers in California navigate this regulatory nightmare? And how do we influence the legislators, engineering board, and others to nudge the statute in the right direction?

California's answer included forming a coalition of professional engineering societies that work together to present a united voice. The California Legislative Council of Professional Engineers (CLCPE) not only represents the interests of engineering professions to the legislature, but serves as a pro bono resource for the legislature on technical issues as well.

This paper will discuss: (1) our lessons learned from working with legislators; (2) California's professional engineering statute and why it is written as it is; and (3) why one needs to be engaged in the legislative process. Professional engineers are entrusted with protecting the health, safety, and welfare of the public. Getting engaged has the benefit of letting legislators understand the impact of their decisions and allows engineers to serve the communities where they live, and provide a safer, better community for everyone.

California's Problem

While most other states utilize some form of a generic registration model for engineers, California's model is discipline-specific, with civil engineering held on a pedestal. The definition of "civil engineering" in the Professional Engineers Act (California Business and Professions Code Sections 6700-6799.) is so broad that it encompasses nearly the entire spectrum of engineering. This overly broad definition, coupled with a mandate that only civil engineers can do civil engineering, means there is little room for those in other engineering disciplines to legally work in their own professions. The law allows overlap from civil engineering into all other disciplines, but other disciplines cannot overlap into civil engineering. So, not only are chemical engineers impacted, the other engineering disciplines are impacted as well.

Section 6731 states the following, with parts underlined for emphasis:

> "Civil engineering embraces the following studies or activities in connection with fixed works for irrigation, drainage, waterpower, water supply, flood control, inland waterways, harbors, municipal improvements, railroads, highways, tunnels, airports and airways, purification of water, sewerage, refuse disposal, foundations, grading, framed and homogeneous structures, buildings, or bridges:
>
> (a) The economics of, the use and design of, materials of construction and the determination of their physical qualities.
>
> (b) The supervision of the construction of engineering structures.
>
> (c) The investigation of the laws, phenomena and forces of nature.
>
> (d) Appraisals or valuations.
>
> (e) The preparation or submission of designs, plans and specifications and engineering reports.
>
> (f) Coordination of the work of professional, technical, or special consultants.
>
> (g) Creation, preparation, or modification of electronic or computerized data in the performance of the activities described in subdivisions (a) through (f)."

For many years the conventional wisdom has been that the term "fixed works," as used in the statute, (see Section 6731 above) somehow limits the scope of civil engineering, leaving room for other engineers to practice their craft. In order to clarify where that boundary lies, Senator Mimi Walters asked the California Legislative Counsel Bureau for a legal opinion (2011). The findings were shocking. "Fixed works" includes not only the structures, but also all the attachments to the structure. So, if a mechanical engineer designs a pumping system or a chemical engineer designs a fractionating column, and that hardware and the hardware is secured to a foundation, those engineers have crossed into the exclusive domain of the civil engineer. Other engineers are similarly impacted.

When a reasonable person reviews this list of studies or activities, it becomes apparent that the list precludes even scientists from legally studying the phenomena and forces of nature. And in most jobs, engineers are expected to produce engineering reports or to coordinate the work of other professionals.

How Did California Get Into This PE Mess?

The perverse nature of the current Professional Engineers Act in California is a result of good intentions gone awry.

In March 1928, a dam northeast of Castaic, CA, broke and flooded the valley below, wiping out the town of Santa Paula and flooding other towns downstream. Hundreds of people died and millions of dollars' worth of property was damaged.

The Professional Engineers Act was then created to ensure that only qualified people (civil engineer PEs) could work on dams and other fixed works. The legislators made the law as inclusive as possible to make sure there were no loopholes. Later, the Legislature expanded the Act to include Chemical, Electrical, Mechanical, Petroleum, Industrial and Metallurgical Engineering as Title Act Disciplines. With the passage of time, the Board for Professional Engineers, Land Surveyors, and Geologists (BPELSG, aka "Engineering Board") was empowered by the Legislature to create Title Disciplines on its own authority, which it did, eventually increasing the count to 15 title protected disciplines (Title Act disciplines), These are defined by Board Rules (16CCR 400-476) and have minimal stature in law. Along the line, Mechanical and Electrical Engineering were converted by statute to practice protection (Practice Act discipline)[1].

This patch job to add mechanical engineering and electrical engineering to the Professional Engineers Act has been ineffective (per Legislative Counsel's review). Inclusion of Title Act disciplines constitutes an even lower tier where those engineers are not able to take responsible charge of work within their field of expertise. In summary, the field is far from level.

The problem with the Professional Engineers Act has been known for many years. The difference now is the Legislative Counsel's legal opinion that was issued in July 2011 (www.CLCPE.org). According to that opinion, companies that rely on engineering services are on notice that the only engineering discipline allowed to conduct engineering work (not just supervise) in California is civil engineering. If a project requires a chemical engineer, a dual-licensed engineer will be needed to meet the competency requirements of both a chemical engineer and the legal requirements of a civil engineer. A dual license is needed for the other engineering disciplines as well.

---

[1] Four of the title disciplines (Corrosion, Manufacturing, Quality and Safety) were dropped ~1999 for lack of an NCEES exam.

The Industrial Exemption

Common to most states, California has an Industrial Exemption that allows companies engaged in manufacturing to use unlicensed engineers in their businesses— except for civil engineering work. The justification for the Industrial Exemption has been that such employers are capable of and competent in evaluating their own engineering talent, that they do not need interference from the state in doing so, and that failure that will impact the public is subject to strict liability.

There are exemptions for people who perform some specific and narrow civil engineering services. Architects performing architecture and building contractors following building codes are examples. But no general exemption exists for engineers who are not civil engineers if they perform "civil engineering." The Industrial Exemption even states that it does not apply to civil engineering.

With the recent legal opinion, however, all engineering practice in connection with facilities that are geographically fixed is now regulated as civil engineering, and the Industrial Exemption is therefore meaningless.

Engaging the Legislature

In the late 1940's, engineering societies in California organized the California Legislative Council for Professional Engineers (CLCPE) in order to work together as a resource to assist the California legislature, support laws that benefit the public and profession, and help keep society members informed about legislation that might affect them. Application of the Professional Engineers Act was not a problem to engineers in the state so long as the Engineering Board and regulators maintained a liberal interpretation. Likewise, CLCPE did not intensify its effort to change the law until the literal interpretation ultimately reinforced by the Legislative Counsel ruling came into focus.

CLCPE's efforts to change the law began in the 1980s. Our members helped draft a bill and found a senator to carry it. In California, the legislator who carries the bill is the "author" and organizations that support the bill, such as CLCPE, are "sponsors" regardless of who actually drafts the bill. Drafting a bill and finding "authors" is a process CLCPE repeated about a dozen times.

Let's reframe the process we used in California in terms that may help others take advantage of our lessons learned.

1) Coalition. We had a consortium of engineering societies working together as a coalition. This provided more credibility in that we were able to speak with one voice instead of individual organizations promoting separate proposals. A coalition also allows a sponsoring organization such as CLCPE to represent a larger portion of a legislator's constituency. AIChE local section leaders met with a senator to promote the one bill that chemical engineers sponsored without going through CLCPE. The senator listened intently about the impact of current law, asked a few clarifying questions, and then asked about how many chemical engineers were in the state. She was not satisfied with our answer, stood up and thanked us for our time.

Creating a coalition of several engineering societies enabled us to afford a lobbyist. AIChE and local sections are a 501(c)(3) nonprofit organization, which restricts how much money can be spent on lobbying. CLCPE however, was organized as a 501(c)(6) nonprofit, which allows for lobbying, but since the societies and CLCPE are all volunteer organizations, we do not have an intense political presence, and that restricts campaigning for or donating to candidates or political parties. Unfortunately, the opposition we face in California has major campaign funding and they openly support candidates that vote as the opposition wants them to vote.

Another advantage of working within a coalition of engineering societies goes beyond promoting legislation. The conversations we had when we met allowed us to gain insight into the problems each of the different disciplines were facing and we heard personal testimony about those various issues. This perspective has been invaluable.

2) Hiring a lobbyist. Over its lifetime, CLCPE has successively employed three lobbyists who helped with efforts to change the Professional Engineers Act, in addition to supporting the overall mission. A good lobbyist helps focus the discussion and facilitate consensus on language for a given bill. They also know how to navigate through the legislature, explain the bill or other proposals to legislators, legislative staff, and to others, and to find legislators who will author a bill. Typically engineers have full-time jobs, and working a bill is very time-consuming. The first of CLCPE's lobbyists was an engineer who later became Executive Officer of the Engineering Board. The other two were professional lobbyists with legal backgrounds.

3) Language for the bill. Earlier this paper discussed drafting language for a bill and then entering that bill into the legislative process (i.e., finding an author). "Entering" is the key word here. In the late 1990s, our bill attracted attention from the Professional and Technical Consultants Association (PATCA). In our quest to expand the law to allow engineers to practice their own profession, we needed to be careful not to adversely impact scientists (e.g., chemists) and other technical professionals. After several meetings with PATCA, we were able to work out compatible language. Going forward on subsequent bills, we deliberately engaged with the chemists (PATCA) early in the process.

4) Bill's "author". Over the decades we have been pursuing change, our bills have been authored by different senators and assemblymen. In several cases, the legislator had been an engineer in his/her former career. One such senator, who had been an aeronautical engineer, had commented that it took multiple disciplines to design an aircraft. A legislator with a technical or science background readily grasped the issue we were dealing with. One of the first committees to hear a bill dealing with engineering (i.e., the bills we were promoting) was either the Business and

Professions Committee (B&PC) in the Senate or the B&PC in the Assembly. When seeking an author we often approached members of these committees as well. The intent was to get early support and move the bill through the legislative process.

5) *Finding support*. In addition to sponsors, it helps to find other organizations that will support a bill through (1) letters, (2) testimonies, and (3) lobbying efforts. We collected binders of letters from individuals, municipalities, regulators, and corporations. These included Chevron, the Chemical Industry Council of California, and Genentech. The letters provided a lot of anecdotes on how this perverse law affected people. The volume of the letters attracted the attention of legislators.

Once a bill is scheduled for a vote, testimonies at the hearing can be very compelling. The city of San Diego, California Fire Prevention Officers Association, a division of the California Fire Chief Association, along with various individual department fire chiefs and fire marshals, individuals impacted by current law, and others have testified at legislative hearings over the years. The Dean of Engineering at University of California—Davis testified on behalf of the University of California at the hearing in January 2012, stating that the state is educating engineers who cannot work in their own professions. The dean was shocked to find out that the legislators still would not vote to support changing the current law.[2]

The third level of support is when companies/ organizations involve their lobbyists. However, we found it very difficult to convince companies that had lobbyists in Sacramento to get engaged at this higher level.

6) *Other curve balls*. We experienced numerous unexpected challenges. One such curve ball happened in 1997, when chemical engineers tried independently (without CLCPE) to change the law. The Engineering Board required that we provide a Sunrise Package[3]. Essentially converting engineers from Title Act (defined in by BPELSG Board Rules) to Practice Act (defined in the Professional Engineers Act) is viewed as creating a new engineering discipline. We were given one week to prepare this. Fortunately, CLCPE had prepared something similar 10 to 15 years earlier, which we were able to mine and then create the required Sunrise Package. The Engineering Board still did not support our bill. Note: Keeping a good archive of material produced in supporting legislation could have unexpected value in the future. The key is being able to find the needed archived information when it is needed. Another advantage is that many of the same people remained engaged in CLCPE for long periods.

---

[2] Incidentally, the western region of ASCE sent their lobbyist to testify in opposition at the same hearing where the UCD dean testified.

[3] Sunrise is a process under which an occupation or profession wishing to receive state certification or licensure must propose the components of the legislation, along with cost and benefit estimates of the proposed regulation

What We Proposed and Why

Our approach to changing legislation evolved over the years. As mentioned earlier, chemical engineers made an attempt in 1997 to raise chemical engineering to a Practice Act (defined by statute). Our opposition protested that our attempt was part of a turf battle among the engineering disciplines. Legislators do not want to be drawn into controversies (e.g., infighting). Our next attempt involved CLCPE, with the intent to define all the Title Act disciplines as Practice Act disciplines. Because we had a coalition involving more than 15 engineering societies, we were able to combat this accusation. Still we did not succeed.

We also shifted our approach and proposed legislation that was closer in alignment with generic engineering. Specifically, we proposed that engineers should be allowed to do work based on their education, experience, or testing. In talking to legislators and to our lobbyist, we learned that we needed to approach any change in engineering statute with baby steps. Adopting generic licensure language (similar to other states or similar to Model Law proposed by the National Society of Professional Engineers (NSPE) was too much at one time. Because of the heavy influence by civil engineering companies in the state, we drafted bills that would continue all the advantages afforded to civil engineering. Unfortunately, our opposition would not negotiate.

In 2000, CLCPE had caught the attention of the legislature sufficient enough that they commissioned a study of the two-tier approach (Title Act/Practice Act) used in California. The study considered natural overlap between the professions, whether there was enough distinction between the engineering disciplines to support this unique regulatory structure, and whether the three Practice Act disciplines (civil, mechanical, and electrical engineering) posed more of a risk to the public (relative to the other engineering disciplines) and thus justified the existing two-tier structure. (See **www.CLCPE.org** for study by the Institute for Social Research.)

The study strongly supported changing the existing regulatory structure. The task of making this change fell to the Engineering Board. The bill it sponsored was opposed by American Consulting Engineers Council (ACEC) - California (a contractors' association) and Professional Engineers in California Government (PECG, a public employee union). Both were well-funded with strong ties to legislators — and the bill failed. [PECG is allowed to donate directly to candidates and its 13,000 members individually campaign for legislators. That has a big influence on those legislators who might support our position.]

The study marked a shift in how we approach changing legislation; CLCPE did not sponsor bills after the study. Instead, CLCPE and our individual engineering societies supported a succession of bills that were sponsored by the Engineering Board or by companies and organizations that use engineering services.

The disadvantage of this approach is that we had less influence on making amendments to the language in the bills. So while sponsoring our own bills had not been successful, this other approach allowed us to continue to pursue fixing the Professional Engineers Act.

Another challenge we had was finding companies and organizations that would commit their lobbyists. Many of these companies were very willing to prepare letters of support, but were not ready to involve their lobbyists.

End Results

This paper provides numerous lessons learned from decades of experience in trying to change California law to allow all engineers to practice in their own professions without finding a civil (typically) or other Practice Act engineer to take responsible charge of the work. Despite years of effort, we have not succeeded in changing the Professional Engineers Act. We are debating our next steps. It is our hope that the lessons we learned will be helpful to others who find the need to engage their state legislators.

(79a) Scaling Biofuels Technology

David Edwards, Sales & Marketing, Zeton Inc, Burlington, ON, Canada

The recommended approach for scaling bioenergy and biofuels technology follows a similar stage gate process to that used in traditional Chemical Process Industry (CPI) processes. But the processing of solids in bioenergy applications leads to some subtle differences, and unique challenges, that need to be considered.

(79b) Challenges in Guard Bed Design for a Biofuel Hydrotreater

Maoqi Feng, Div. of Chemistry and Chemical Engineering, Southwest Research Institute, San Antonio, TX

Bio-oils derived from fragmentation of biomass building blocks (cellulose, hemicellulose, and lignin) are unstable. Hydrotreatment of the oxygenated bio-oils to drop-in hydrocarbon fuels is necessary. Traditional hydrotreating catalysts, such as sulfided CoMo or NiMo supported on alumina have been reported in hydrodeoxygenation of biofuels and the process conditions are similar to those used in the refining of petroleum cuts. Besides the high oxygen content in bio-oils, bio-oils also differ from petroleum feedstock in sulfur, phosphorus and metal content. To prevent the hydrotreating catalyst fast deactivation and prolong the catalyst lifetime, a guard bed is necessary, and the guard bed design is very important. This presentation will address some challenging issues in the guard bed design for bio-oils hydrotreating.

(79c) Non-Geometric Scale-up to Meet Mixing Requirements

David S. Dickey, MixTech, Inc., Dayton, OH

Scale-up is a term most often used for mixing processes, because mixing is about both process results and fluid flow patterns. The interaction between process results and flow patterns can be a complicated relationship. Simple rules apply to some applications and are often combinations of factors related to both fluid motion and mixing intensity. As mixer geometry changes with different vessel sizes and impeller types, flow patterns change. As flow patterns change, so does mixing intensity and often process results.

Keeping important mixing parameters the same in different size vessels requires maintaining some similarities in flow patterns and mixing intensity to achieve comparable process results. As mixer and vessel geometry changes between the pilot scale and the production equipment, adjustments need to be made in the mixing behavior.

Geometric similarity provides a convenient way of achieving similar flow patterns, but large scale stirred tanks rarely have the same geometry as the pilot plant equipment. Finding ways of maintaining similar flow patterns and mixing intensity in not only different size equipment, but with different tank geometries and impeller types requires multiple adjustments to achieve desired results.

A systematic scale-up method combining geometric similarity and non-geometric adjustments provides a step-by-step process to observe and confirm changes in mixing intensity and flow patterns. No single method always works, but the chances of success increase when the effects of changes can be estimated and observed by basic mixing calculations. A multi-step method for non-geometric mixing scale-up will be presented.

(79d) Polycondensation Process System Design and Scale-up Challenges

Frank P. Lee, Xerox Research Centre of Canada, Mississauga, ON, Canada

Polycondensation reactions have many industrial applications and are commonly employed in syntheses of polymers such as polyesters. A typical Polycondensation process takes place at relatively high temperature and can involve exothermic and/or endothermic reactions. Small molecules are formed by the condensation reaction and are usually removed by distillation in order to drive the reaction to completion and to isolate the products. Once the small molecules are removed, the bulk of the product can become very viscous even at high temperature. The viscosity of molten raw materials and products can range from water-like to more than 10,000 Centipoises.

An institution actively engaged in development and scale up of specialty polymer can benefit from having versatile Bench Engineering and Pilot Plant process systems custom designed and built specifically for carrying out Polycondensation. A typical Polycondensation reaction system should incorporate the following components and features:

- A reactor which can handle a broad range of mixing (low to very high viscosity materials) and heat transfer (exotherm and endotherm) requirements, as well as having adequate numbers of nozzles sized for feeding bulky raw materials and discharging viscous products

- Auxiliary equipment including:

High-temperature, high-power circulation baths for the reactor and condensers
Refrigerated water or glycol supply system(s) for condensers and receivers

Atmospheric, or preferably, vacuum distillation equipment such as reflux and take-off condensers, packed columns, vacuum pump, splitter and phase separation equipment

Vessels and pumps for feeding and collecting raw materials, products and byproducts

- Process Monitoring, Control and Data Acquisition components and systems for monitoring and control of process parameters, including temperature, pressure, viscosity/torque/power, flow and agitator speeds

- Sampling and analytical instruments for end point determination

- Post reaction processing equipment such as chilled rolls or belts

In this presentation, examples of several custom Bench Engineering and Pilot Plant process system, ranging in size from less than 1 litre to over 1,000 litres, which were designed and built specifically for carrying out a broad range of Polycondensation reactions in a materials research centre are described. Challenges and lessons learned from designing, operating and scale-up of Polycondensation processes using these custom reaction systems should interest those who are involved with development and scale-up of polymer syntheses.

(79e) Pilot Plants in Transition

Mike J. Doll, HODer, Shell Global Solutions, Houston, TX

We, as chemical engineers, discuss pilot plant design, construction, operation, and management in great detail. However, we seldom, if ever, discuss the pilot plant in transition: that is --- we never discuss moving pilot plants from one location to another location.

This presentation discusses the reasons for moving a pilot plant, how to move a pilot plant safely and expeditiously, how to rebuild it and how to safely restart it. This presentation also discusses designing pilot plants with the view that they may be moved at some future date.

On the Sensitivity of "Overshoot" to Initial and Boundary Conditions and Engineering and Mathematical Perpsectives

ABSTRACT:

Output response of "temperature overshoot"" during transient non-Fourtier, heat conduction in a finite slab in response to a step change at the boundary is studied systematically. Why the overshoot appears in some systems and why it disappeares in others is explained thoroughly. A conclusion is drawn that the "overshoot pheonemena" is more a mathematical artifact rather than depiction of transient heat transfer events that can be expected in practice. The initial accumulation time condition was evaluated side by side for values of 0 initial accumulation, when initial accumulation is equal to the dimensionless group –Ve, Venrnote number respectively for a finite slab subject to the constant wall temperature boundary condition. For a material with large relaxation time the overshoot was found for the model results when zero initial accumulation time condition. For the same set of material and parameters when a physically reasonable time accumulation condition was used the overshoot disappears. However, the transient temperature was found to be subcritical damped oscillatory. Lumped analysis was used to study the average temperature in a finite slab subject to convective heating using; (i) Fourier parabolic model. The model solution was found to rise monotonically to a constant asymptotic value; (ii) Hyperbolic model with the first derivative of temperature with respect to time set to zero. Solution was obtained by the method of Laplace transforms. This model solution appears to have an overshoot; (iii) Hyperbolic model with the initial temperature at T_0 and the additional constraint that the average transient temperature should obey the energy balance equation from a engineering lumped analysis. Closed form model solutions were obtained after making the equation homogenous and illustrated in Figures. As shown in Figures 6.0-8.0, the model solution does not exhibit any overshoot when the boundary condition includes a finite wave term contribution.

A dimensionless group, *Sharma Number (heat)* is introduced. This appears to be an important parameter in determination of the average transient temperature in a finite slab during damped wave conduction and relaxation. Expression for the time taken to attain steady state was derived. The maxima in the transient temperature were found to increase with decreasting S* starting with large values such as 10. A cross-over was found after S* became less than about 2.2. Then the maxima in the average transient temperature was found to decrease with decrease in Sharma Number (heat), S*. Eq. (62) can be seen to go to zero in the asymptotic limits of infinite Sharma Number (heat), $S^* \rightarrow \infty$.

Keywords: Taitel Paradox; Overhoot Phenomena; Damped Wave Conduction and Relaxation; Wave Term in Boundary Condition; Intial and Final Time Conditions; Engineering Analysis

---

1. INTRODUCTION

There are reports in the literature [1,2] about occurrence of temperature overshoot upon use of non-Fourier heat conduction equation. The "overshoot phenomena" is studied herein a systematic manner. The Taitel paradox [3] is an anamoly seen in the model solution for transient temperature, in a finite slab, heated by isothermal-hot walls, btained by the method of separation of variables using the hyerpbolic PDE, partial differential equation. For certain values of the thermophysical properties of the finite slab the interior temperature of the slab was found to be greater than the wall temperature. This "overshoot" lead practioners to abandon the non-fourier heat conduction equation concluding that the equation violated second law of thermodynamics and hence cannot be used for engineering applications. A careful side by side study of this problem is presented below. The time condition $\frac{\partial T}{\partial t}$ was assumed to be 0 in Taitel [3].

Consider a finite slab of width 2a, at a initial temperature of T_0 heated by hot isothermal walls brought a temperature, $T_s < T_0$ for times $t > 0$, greater than 0. As can be seen from Figure 1.0 below only the *left hand limit* of the rate of change of average temperature of the slab with respect to time is zero. A lumped analysis on the heating process leads to the expression for the rate of change of average temperature of the interior of the slab as shown in the ordinate of Figure 1.0.

$$\frac{\partial <T>}{\partial t} = \frac{\alpha}{a^2} e^{-\frac{\alpha t}{a^2}} \tag{1}$$

The *right hand limit*, of the rate of change of average temperature in the finite slab is a maximum! With increase in time this decreases exponentially to a asymptotic zero at large times or upon attainment of steady state.

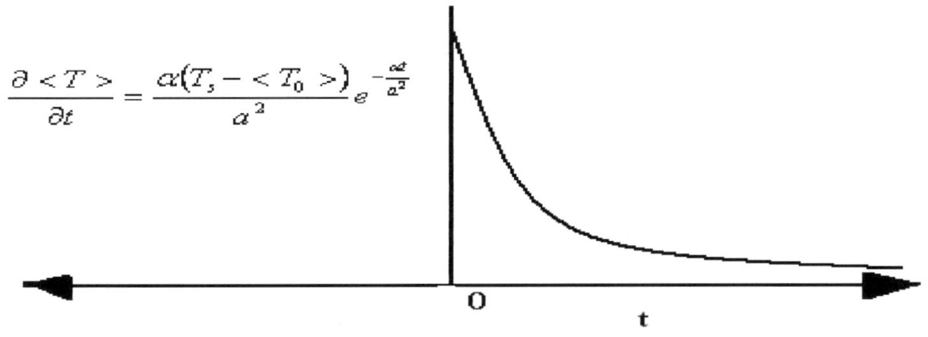

Figure 1.0 Rate of Change of Average Temperature in Interior of Finite Slab as a Function of Time, t

The right hand limit was used in the analysis**.** For a material with large relaxation time the overshoot was found for the model results when initial accumulation time condition is taken as 0. For the same set of material and parameters when a physically reasonable time accumulation condition was used the overshoot disappears. The transient temperature was subcritical damped oscillatory. A steady state temperature was attained after a said time. Lumped analysis was further explored.

The average temperature in a finite slab subject to convective heating was obtained using:

(i) Fourier parabolic model. The model solution rises monotonically to a constant asymptotic value.

(ii) Hyperbolic model with the first derivative of temperature with respect to time set to zero as u by the method of Laplace transforms. This model solution appears to have an overshoot;

(iii) Hyperbolic model with the initial temperature at T_0 and the additional constraint that the average transient temperature should obey the energy balance equation from a lumped analysis. The dimensionless temperature was expressed as a sum of a steady state temperature and a transient temperature. The transient temperature was expressed as a product of wave temperature and decaying exponential in time. The model solution was found to be;

$$< u >^{t} = \left(\frac{S^{*}}{\omega(1 + \beta S^{*})} \right) e^{-\left(\frac{\tau}{2}(1+\beta S^{*})\right)} (\sin(\omega\tau))$$

(2)

Where, S* is the dimensionless Sharma Number (heat). As shown in Figure the model solution does not exhibit any overshoot. It appears that the damped wave conduction and relaxation equation can be applied to transient heat conduction problems without violation of the second law of thermodynamics. The Sharma Number (heat) appears to be an important parameter in determination of the average transient temperature in a finite slab during damped wave conduction and relaxation. The time taken to attain steady state was found to be ;

$$\tau_{ss} = \left(\frac{\pi}{\omega} \right)$$

(3)

The maxima in the transient temperature were found to increase with decreasing S* starting with large values such as 10. A cross-over was found after S* became less than about 2.2. Then the maxima in the average transient temperature was found to decrease with decrease in Sharma Number (heat), S*. The average transient temperature becomes zero in the infinite limit of $S^* \rightarrow \infty$.

2. LITERATURE REVIEW

The damped wave conuction and relaxation equation was sought over Fourier's law of heat conduction for eight reasons by Sharma [4]. These reasons include:

(i) contradiction with the principle of microscopic reversibility by nobel laureate, Onsager, [5]);

(ii) neglects the time needed for the acceleration of heat flow by free electrons, Sharma [6];

(iii) occurrence of "blow-up" in the solutions obtained using Fourier parabolic model for industrially important systems such fluidized bed heat transfer [7], CPU overheating [8]), gel electrophoresis, [9], restriction mapping [10], adsorption [11], nuclear fuel rod [12] and drug delivery systems [13] ;

(iv) heat transfer in the Casimir limit [14] or nanoscale systems cannot be described using Fourier's model;

(v) basis for Fourier's law was from empirical observations at steady state;

(vi) oscillatory discharge of heat in good thermal conductors at low temperature as poinetd out by nobel laureate, Nernst [15];

(vii) ultrafast laser heating of metals [16], delayed ignition of solid propellant [17] cannot be explained using Fourier's model;

(viii) speed of heat cannot be greater than the speed of light as discussed by nobel laureate Landau and Lifshitz, [18].

The damped wave conduction and relaxation equation was originally suggested by Maxwell [19], and postulated independently by Cattaneo [20,21] and Vernotte [22]. The damped wave conduction and relaxation equation in one dimension across constant area may be written as follows;

$$q_x = -kA\frac{\partial T}{\partial x} - \tau_r A\frac{\partial q_x}{\partial t} \qquad (4)$$

Where q_x is the heat transfer rate in x direction in (watts, w), A is the cross-sectional area across which the heat conduction occurs in (m^2), k is the thermal conductivity of the material in ($w.m^{-1}.K^{-1}$), τ_r is the relaxation time (s).

Reviews of the use of this equation have been presented by Joseph and Preziosi [23,24] and Ozisik and Tzou [25]. Extensive theoretical treatment of the equation have been reported by Tzou [26] and Sharma [27]. Experimental measurement of relaxation times has been reported by Mitra et. al. [28] recently for biological materials. Taitel [3] found a overshoot in his transient temperature solution for a finite slab subject to constant wall temperature boundary condition. Bai and Lavine [29] was concerned about Eq. (4) violating the second law of thermodynamics. Zanchini [30], Barletta and Zanchini [31], calculate a entropy production term and are concerned of a violation of Clausius' inequality. Al Nimir et. al. [32,33] discuss an "overshoot" and equilibrium entropy production. Haji Sheik et. al. [34] point out some anomalies in Eq. (4). Tzou [26] has found Eq. (3.68) to be admissible within the framework of second law of thermodynamics. Sharma [27,36-40] have presented closed form analytical solutions for different geometries that are within the bounds of second law of thermodynamics. He has derived the damped wave equation by accounting for acceleration of the molecules in the Stokes-Einstein formulation [41]. Antaki [42] has discussed some analytical solutions for convective boundary condition.

Why the "overshoot" manifests in such systems from the model development and implications on the physical insight to the phenomena are discussed in the following subsection. The use of physically realizable time condition such as obeyance of hyperbolic transient temperature to the energy balance equation from a lumped analysis approach can lead to the *disappearance* of the overshoot is shown . The time condition used in this analysis is *de novo*.

3. MATHEMATICAL MODEL

3.1.1 REVISIT OF TAITEL PARADOX

Taitel [3] considered a finite slab (Figure 1.0) with two boundaries of width 2a heated from both sides. Both the sides are maintained at a constant temperature Ts for times, t > 0. At initial time, t = 0, the temperature at all points in the slab is T_0. The governing equation is given by Eq. (5).

$$\frac{\partial u}{\partial \tau} + \frac{\partial^2 u}{\partial \tau^2} = \frac{\partial^2 u}{\partial X^2} \tag{5}$$

Where u is given as $u = \left(\frac{T - T_s}{T_0 - T_s} \right)$

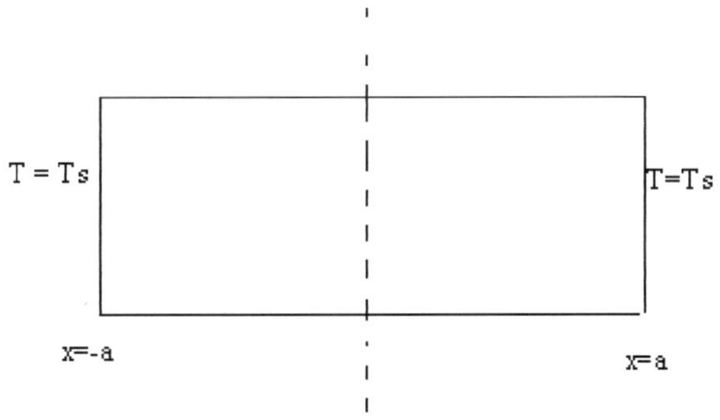

Figure 2.0 Finite Slab with Two Boundaries Maintained at Constant Temperature, Ts

The four conditions used by Taitel, two in space and two in time that are needed to completely describe a hyperbolic PDE that is second order with respect to space and with respect to time were;

$$t = 0,\ -a < x < +a,\ T = T_0, \text{ or } u = 1 \tag{6}$$

$$t > 0,\ x = \pm a,\ T = T_s,\ u = 0 \tag{7}$$

$$t = 0, \quad \frac{\partial u}{\partial \tau} = 0 \tag{8}$$

Taitel presented a solution for Eq. (5) for the conditions stated above obtained by analytical methods in a finite slab. The solution obtained by Taitel, for the centerline temperature of the finite slab is given below. They considered a constant wall temperature and the initial time conditions included a $\frac{\partial T}{\partial t} = 0$ term in addition to the initial temperature condition. They found that for certain values of the parameters used, i.e., relaxation time, width of the slab the transient temperature within the slab was found the exceed the wall temperature. Sharma [9] used a final condition in time and developed well bounded solutions . For materials with large relaxation times, the transient temperature exhibited subcritical damped oscillations. Multiplying throughout Eq. (2) by exp(nτ); Let $w = ue^{n\tau}$. Then,

$$\frac{\partial w}{\partial \tau} = e^{n\tau}\frac{\partial u}{\partial \tau} + ne^{n\tau}u = nw + e^{n\tau}\frac{\partial u}{\partial \tau}$$
$$\frac{\partial^2 w}{\partial \tau^2} = n\frac{\partial w}{\partial \tau} + ne^{n\tau}\frac{\partial u}{\partial \tau} + e^{n\tau}\frac{\partial^2 u}{\partial \tau^2} \tag{9}$$

Combining Eq. (9) and Eq. (5),

$$\frac{\partial^2 w}{\partial X^2} = \frac{\partial w}{\partial \tau} - nw + \frac{\partial^2 w}{\partial \tau^2} - 2n\frac{\partial w}{\partial \tau} + n^2 w \tag{10}$$

For n = ½, Eq. (5) becomes;

$$\frac{\partial^2 w}{\partial X^2} = \frac{\partial^2 w}{\partial \tau^2} - \frac{w}{4} \tag{11}$$

w in Eq. (9) is the wave temperature. Eq. (9) can be solved by the method of separation of variables.

$$\text{Let } u = V(\tau)g(X) \tag{12}$$

Eq.(9) becomes,

$$\frac{g''(X)}{g(X))} = \frac{V''(\tau)}{V(\tau)} - \frac{1}{4} = -\lambda_n^2 \tag{13}$$

$$g(X) = c_1 \mathrm{Sin}(\lambda_n X) + c_2 \mathrm{Cos}(\lambda_n X) \tag{14}$$

From the boundary conditions,

$$\text{At } X = 0,\ \partial g/\partial X = 0,\ \text{So, } c_1 = 0 \tag{15}$$

$$\phi(X) = c_1 \mathrm{Cos}(\lambda_n X) \tag{16}$$

$$0 = c_1 \mathrm{Cos}(\lambda_n X_a) \tag{17}$$

$$\lambda_n = \left(\frac{2n-1}{2a}\right)\left(\frac{\pi\sqrt{\alpha\tau_r}}{2}\right) \tag{18}$$

$n = 1, 2, 3 \ldots\ldots$

The time domain solution would be, for, $\lambda_n \leq 0.5$,

$$V = e^{-\frac{\tau}{2}}\left(c_3 e^{\tau\sqrt{0.25-\lambda_n^2}} + c_4 e^{-\tau\sqrt{0.25-\lambda_n^2}}\right) \tag{19}$$

and

$$\omega_n = \sqrt{\lambda_n^2 - 0.25}$$

The dimensionless temperature can be seen to be for materials with large relaxation times when Eq. (19) is applicable is given by an infinite series as follows;

$$u = \sum_{n=1}^{\infty} e^{-\frac{\tau}{2}} \cos(\lambda_n X)(c_n \cos(\omega_n \tau) + d_n \sin(\omega_n \tau)) \tag{20}$$

For the set of time conditions given in Taitel and by Eqs. (3,5) the cn and dn were solved for by use of principle of orthogonality and found to be;

$$c_n = \frac{4(-1)^{n+1}}{(2n-1)\pi}; d_n = \frac{2(-1)^{n-1}}{(2n-1)\pi} \tag{21}$$

The transient temperature u was computed as a function of time using MS Excel for windows 2007 spreadsheet in a desktop computer with 1.8 GHz microprocessor speed. 12 Terms in the infinite series in Eq. (20) were taken for a material with a relaxation time, $\tau r = 1.4$ s at dimensionless distance X = 0.1; half-width a = 5 mm; thermal diffusivity, $\alpha = 10^{-5}$ $m^2.s^{-1}$. The Vernote number corresponding to this simulation was;

$$Ve = \frac{\alpha \tau_r}{a^2} = 0.56 \tag{22}$$

The results of u vs. τ for the above set of parameters is shown in Figure 2.0. An overshoot can be seen.

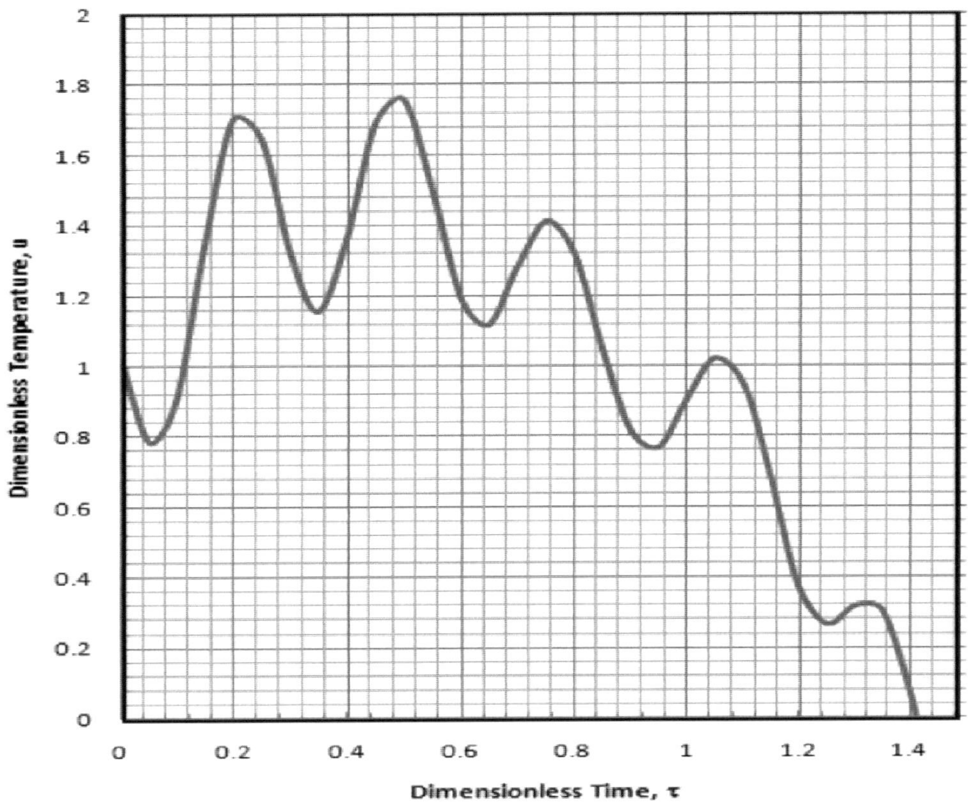

Figure 3.0 Transient Temperature inside the Finite Slab with Time Condition = 0
for Parameters: Location X = 0.1; Relaxation Time τ_r = 1.4s;Half-width of Slab, a = 5 mm; Thermal Diffusivity, $\alpha = 10^{-5}$ **m2.s^{-1}**

3.1.2 ACCUMULATION TIME CONDITION FROM LUMPED ANALYSIS AND ENERGY BALANCE

Examination of Figure 1.0 indicates that the value 0 for the initial accumulation of temperature in time is the "left hand limit". At t =0 the two sides of the slab abruptly brought to a higher temperature, $T_s > T_0$ where, T_0 is the initial slab temperature. The right hand limit may be more appropriate for purposes of obtaining exact closed form analytical solutions. Consider a lumped analysis as follows; Let the entire temperature of the slab other than the surfaces be lumped into an average temperature <T>. Let this be assumed to be at the center of the slab (centroid). Then in order to keep the surface temperature at T_s energy needs to be supplied from the surroundings. In turn the interior of the slab is heated by the hot surfaces. Consider the finite slab less the surface as the control volume. The energy balance can be written as follows;

$$\rho VC_P\left(\frac{\partial <T>}{\partial t}\right) = kA\left(\frac{T_s - <T>}{a}\right)$$

$$\frac{d<u>}{dt} = -Ve<u>$$

(23)

Where, Vernotte number, Ve is given by . $Ve = \left(\frac{\alpha\tau_r}{a^2}\right)$. This is assuming that Volume to Area ratio is given by the half-width a. So the "right hand limit" of the initial accumulation time condition closer to Eq. (23). For time, t =0, <u> = 1. Hence a more realistic time condition would be;

$$\frac{\partial u}{dt} = -Ve$$

(24)

Eq. (24) was used instead of Eq. (5) in the second time condition. The dimensionless temperature remains the same as given in Eq. (20) and is;

$$u = \sum_{n=1}^{\infty} e^{-\frac{\tau}{2}} \cos(\lambda_n X)(c_n \cos(\omega_n\tau) + d_n \sin(\omega_n\tau))$$

(25)

The constants cn and dn are recalculated for the time accumulation condition given by Eq. (23) and found to be;

$$c_n = \frac{4(-1)^{n+1}}{(2n-1)\pi}; d_n = (1-2Ve)\frac{2(-1)^{n-1}}{(2n-1)\pi}$$

(26)

The transient temperature u was computed as a function of time using MS Excel for windows 2007 spreadsheet in a desktop computer with 1.8 GHz microprocessor speed. 12 Terms in the infinite series in Eq. (26) were taken for the same material with a relaxation time, $\tau_r = 1.4$ s at dimensionless distance X = 0.1; half-width a = 5 mm; thermal diffusivity, α = 10-5 $m^2.s^{-1}$. The Vernote number corresponding to this simulation was;

$$Ve = \frac{\alpha \tau_r}{a^2} = 0.56$$

The results of u vs. τ for the above set of parameters is shown in Figure 3.0. An overshoot cannot be seen! The oscillations in temperature are subcritical. The transient temperature reaches a steady state value! after a said time.

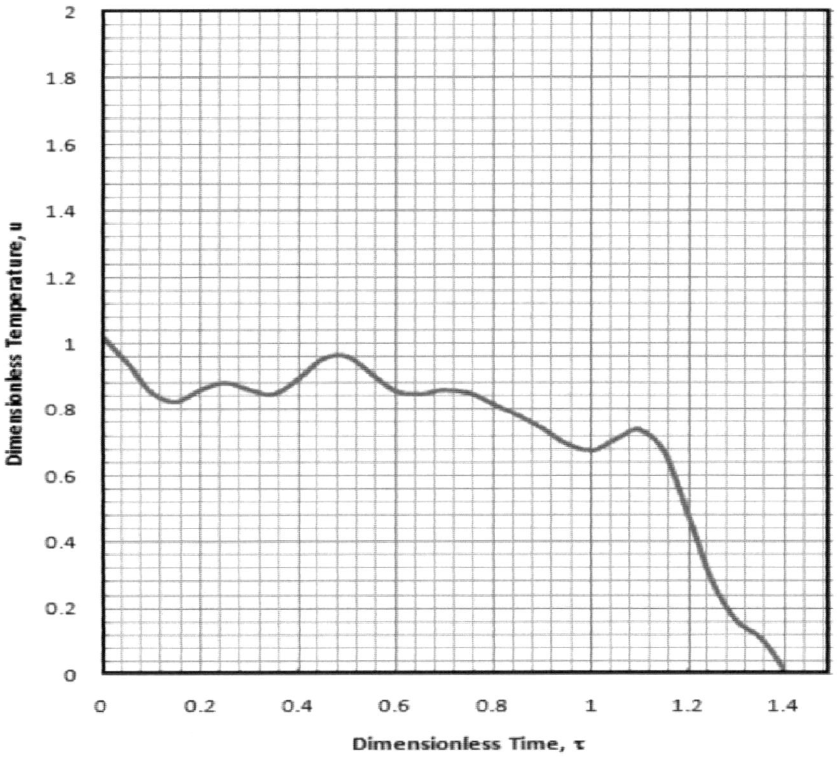

Figure 4.0 Transient Temperature inside the Finite Slab with Time Condition *Ve*for Parameters: Location X = 0.1; Relaxation Time τ_r = 1.4s;Half-width of Slab, a = 5 mm; Thermal Diffusivity, $\alpha = 10^{-5}$ $m^2.s^{-1}$

3.2 AVERAGE TEMPERATURE IN A FINITE SLAB

3.2.1 LUMPED/ENGINEERING ANALYSIS

Consider a finite slab with width 2a at a initial hot temperature T_h. This hot slab part of which is shown in Figure 5.0 is subject to a sudden cooling by convection by the fluid blown from the fan as shown in Figure 4 for times t > 0. The fluid temperature is at T_f where $T_f < T_h$. Both sides of the slab are cooled by the fluid. The transient temperature as a function of space and time need be obtained. The two time and two space conditions are given as follows;

$$t = 0,\ -a \le x \le +a,\ T = T_h \tag{27}$$

$$t > 0,\ x = 0,\ \frac{\partial T}{\partial x} = 0 \tag{28}$$

$$t > 0,\ x = \pm a,\ -k\frac{\partial T}{\partial x} = h\left(T - T_f\right) \tag{29}$$

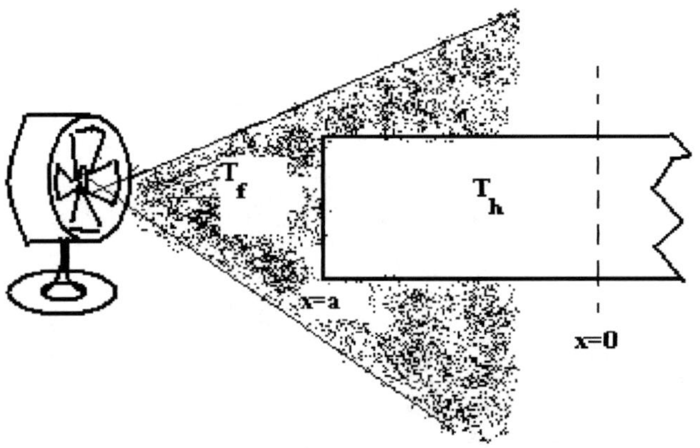

Figure 5.0 Finite Slab with Half-Width a Subject to Convective Heating

In one dimension the energy balance equation on a thin slice of thickness Δx can be written as;

$$-\frac{\partial q}{\partial x} = \left(\rho C_p\right)\frac{\partial T}{\partial t} \tag{30}$$

Combining Eq. (30) with the damped wave heat conduction and relaxation equation the governing equation can be written as;

$$\frac{\partial u}{\partial \tau} + \frac{\partial^2 u}{\partial \tau^2} = \frac{\partial^2 u}{\partial X^2} \tag{31}$$

$$\text{where, } u = \frac{(T - T_h)}{(T_f - T_h)}; \tau = \frac{t}{\tau_r}; X = \frac{x}{\sqrt{\alpha \tau_r}} \tag{32}$$

The time and space conditions are as follows;

$$\tau = 0, \quad u = 0 \tag{33}$$

$$X = 0, \quad \frac{\partial u}{\partial X} = 0$$

$$X = \frac{a}{\sqrt{\alpha \tau_r}}$$
$$-h^*(u - 1) = \frac{\partial u}{\partial X} \tag{34}$$

$$\text{Where, } h^* = \left(\frac{h}{\sqrt{\frac{k \rho C_p}{\tau_r}}} \right)$$

It can be noted that the heat transfer coefficient is defined with respect to the average temperature in the slab <T>. The energy balance on a thin spherical shell at x with thickness Δx is written. The governing equation can be obtained after eliminating q between the energy balance equation and the derivative with respect to x of the flux equation and introducing the dimensionless variables;

$$\frac{\partial u}{\partial \tau} + \frac{\partial^2 u}{\partial \tau^2} = \frac{\partial^2 u}{\partial X^2} \tag{35}$$

Eq. (35) is integrated between $-X_a$ and $+X_a$ and give;

$$\frac{\partial <u>}{\partial \tau} + \frac{\partial^2 <u>}{\partial \tau^2} = -S^*(<u> -1) \tag{36}$$

Where the Sharma Number (heat) S^* is defined as $S^* = \left(\frac{h}{Sa}\right) = \left(\frac{h\tau_r}{\rho C_p a}\right)$.

S is the storage number with units of $(w.m^{-3}.K^{-1})$ and is given by;

$$S = \left(\frac{\rho C_p}{\tau_r}\right)$$

The physical significane of storage number S is that is a material property such as thermal conductivity with units of $(w.m^{-1}.K^{-1})$ and system parameter such as convective heat transfer coefficient, h with units of $(w.m^{-2}.K^{-1})$. Storage number is a measure of energy stored in the material. The use of storage number is illustrated in another study on CPU overheating [8]. It is the ratio of the thermal mass of the material to its relaxation time. What is the material's thermal capacitance in relation to the speed of heat conduction is captured in this one number. Sharma Number (heat) is the dimensionless ratio of the convective heat transfer to the wave heat propagation heat transfer and storage in a finite region a.

3.2.2 FOURIER PARABOLIC MODEL NEAR STEADY STATE

Eq. (35) in the limit of zero relaxation time reverts to the Fourier parabolic model and the governing equation becomes;

$$\frac{\partial <u>}{\partial \tau} = -S^* <u> -1) \tag{37}$$

Obtaining the Laplace transforms of Eq. (37) ;

$$s < u > (s) - 0 + S^* < u > (s) = \frac{1}{s} \qquad (38)$$

or,

$$< u > (s) = \frac{1}{s\left(s + S^*\right)} \qquad (39)$$

Obtaining the inverse Laplace transform of Eq. (39);

$$< u > = \int_0^\tau e^{-\frac{hS}{a}p} dp = 1 - e^{-\frac{ht}{\rho C_p a}} \qquad (40)$$

Eq. (40) is the average temperature of the finite slab in response to a step change in convective cooling at both sides of the finite slab. Eq. (40) is not a function of the relaxation time and is a function of the heat transfer coefficient, h, density, ρ, and heat capacity, C_p, and half-width of the finite slab, a.

3.2.3 INITIAL ACCUMULATION CONDITION ZERO

For the hyperbolic model other investigators such as Taitel [3] use the following initial condition;

$$\tau = 0, \quad \frac{\partial u}{\partial \tau} = \frac{\partial < u >}{\partial \tau} = 0 \qquad (41)$$

Obtaining the Laplace transform of Eq. (37) realizing the initial condition from Eq. (41)) the Laplace domain expression for <u> (s) can be seen to be;

$$< u > (s) = \frac{1}{s\left(\frac{s^2}{S^*} + \frac{s}{S^*} + 1\right)} \qquad (42)$$

Where, $S^* = \left(\frac{h}{Sa}\right)$

Eq. (42) can be compared with the prototypical second order systems with a step input [42,43] For underdamped systems, i.e., when the damping factor, $\zeta < 1$ the time response to Eq. (42) can be seen to exhibit a "overshoot". Without *any heat sources* the internal temperature greater than the boundary temperature can be a violation of second law of thermodynamics. When the damping coefficient,

$$\tau_p^2 = \frac{1}{S^*} \tag{43}$$

$$2\zeta \frac{1}{\sqrt{S^*}} = \frac{1}{S^*} \tag{44}$$

So, the damping factor is;

$$\zeta = \frac{1}{2\sqrt{S^*}}$$

For large values of the dimensionless ratio, S^*, the damping coefficient will become less than one and the output response appears to have a overshoot. This can be expected for materials with large relaxation time,

$$\tau_r > \left(\frac{\rho C_p a}{h}\right) \tag{45}$$

3.2.4 PHYSICALLY REALIZABLE STEADY STATE/FINAL TIME CONDITION

The boundary condition used during convective cooling is a constant value for the heat flux rate from the air flowing past the finite slab as given in Eq. (29). The governing equation (35) is the damped wave conduction and relaxation equation taking into account the effect of acceleration of electrons. A boundary condition that includes the wave term, i.e, $\tau_r^{air}\dfrac{\partial q}{\partial t}$ would be a better representation of transient events at the air solid interface. Thus. Eq. (3.141) need be modified as;

$$-\frac{\partial u}{\partial X} = h^*(<u>-1) + h^*\left(\frac{\tau_r^{air}}{\tau_{rs}}\right)\left(\frac{\partial <u>}{\partial \tau}\right) \tag{46}$$

Eq. (35) is integrated with respect to X and the governing equation written in terms of the average temperature in the slab <u> as follows;

$$\frac{\partial <u>}{\partial \tau}(1+\beta S^*) + \frac{\partial^2 <u>}{\partial \tau^2} + S^* <u> = S^* \tag{47}$$

where, $\beta = \left(\dfrac{\tau_r^{air}}{\tau_{rs}}\right)$ the ratio of the relaxation time in the air and the relaxation time in the solid. Let the <u> be expressed as a sum of transient temperature and steady state temperature.

$$<u> = <u>^{ss} + <u>^t \tag{48}$$

Substituting Eq. (48) into Eq. (47);

$$\frac{\partial <u>^t}{\partial \tau}(1+\beta S^*) + \frac{\partial^2 <u>^t}{\partial \tau^2} + S^* <u>^t + <u>^{ss}(1-S^*) = 0 \tag{49}$$

Eq. (49) is valid when the following two equations are valid;

$$(1+\beta S^*)\frac{\partial <u>^t}{\partial \tau} + \frac{\partial^2 <u>^t}{\partial \tau^2} + S^* <u>^t = 0 \tag{50}$$

$$+S^*(<u>^{ss}-1) = 0 \tag{51}$$

The steady state temperature can be seen to be;

$$<u>^{ss} = 1 \tag{52}$$

Eq.(50) is a second order differential equation that has been made homogeneous. Eq. (50) is multiplied by $e^{n\tau}$ throughout. The terms group such that,

$$<W>^{t} = e^{+n\tau} <u>^{t} \quad (53)$$

and for $n = \left(\frac{1+\beta S^{*}}{2}\right)$ Eq. (50) can be seen to become;

$$\frac{\partial^2 <W>^{t}}{\partial \tau^2} = -<W>^{t}\left(\left|S^{*}-1\right|^{2}\right) \quad (54)$$

The solution to Eq. (54) can be seen to be;

$$<W>^{t} = c_1 \cos(\omega\tau) + c_2 \sin(\omega\tau) \quad (55)$$

Where $\omega = \left|S^{*}-1\right|$

Two time conditions are needed to fully describe the problem. The accumulation condition equal to zero appears to be physically unreasonable. When a slab with a initial temperature is immersed in a hot fluid with convective heat transfer coefficient the temperature in the slab would increase, rapidly in the beginning of the process and can be expected to slow down with the passage of time. At steady state the temperature would be steady and invariant with time. From the initial condition given by Eq. (32) the integration constants in Eq. (55) can be seen to be;

$$c_1 = 0 \quad (56)$$

Then Eq.(53) becomes;

$$(<u>^{t} = e^{-\frac{(1+\beta S^{*})\tau}{2}}(c_2 \sin(\omega\tau)) \quad (57)$$

The transient average temperature $<u>^{t}$ also has to obey the following equation that can be obtained from a lumped analysis of the finite slab from energy balance;

$$\frac{\partial < u >^t}{\partial \tau}(1+\beta S^*) = S^*\left(1- < u >^t\right) \quad (58)$$

Eq. (58) at zero time, or τ=0, can be seen to be!

$$\frac{\partial < u >^t}{\partial \tau} = \left(\frac{S^*}{1+\beta S^*}\right) \quad (59)$$

The integration constant c_2 can be obtained by applying Eq. (3.166) to Eq. (3.167).

$$\frac{\partial < u >^t}{\partial \tau} = -\frac{c_2\left(1+\beta S^*\right)}{2} e^{-\left(\frac{\tau}{2}\left(1+\beta S^*\right)\right)} \sin(\omega\tau) + c_2 \omega e^{-\left(\frac{\tau}{2}\right)} \cos(\omega\tau) \quad (60)$$

and c_2 seen to be; $$c_2 = \frac{S^*}{\omega\left(1+\beta S^*\right)} \quad (61)$$

The average transient temperature in the slab is then written as;

$$< u >^t = \left(\frac{S^*}{\omega\left(1+\beta S^*\right)}\right) e^{-\left(\frac{\tau}{2}(1+\beta S^*)\right)} \left(\sin(\omega\tau)\right) \quad (62)$$

Eq. (62) is plotted in Figures 6 – 8, for different values of S^* and β. The time taken for the slab to reach steady state is finite and can be calculated when ($\omega\tau_{ss}$) becomes π in Eq. (62). This happens when;

$$\tau_{ss} = \left(\frac{\pi}{\omega}\right) \quad (63)$$

For the special case when S* = 1, Eq. (54) becomes;

$$\frac{\partial^2 <W>^t}{\partial \tau^2} = 0$$

$$<W>^t = c_1 \tau + c_2 \qquad (64)$$

Or,

$$<u>^t = e^{-\frac{(1+\beta)\tau}{2}} (c_2 + c_1 \tau) \qquad (65)$$

From the initial condition given by Eq.(8), $c_2 = 0$. c_1 can be solved for the energy balance equation given by Eq. (3.166) at $\tau = 0$,

$$\frac{\partial <u>^t}{\partial \tau} = \left(\frac{1}{1+\beta} \right) = -\frac{c_1 (1+\beta)\tau}{2} + c_1 \quad ; \quad c_1 = \frac{1}{(1+\beta)}.$$

Thus for the special case when S*=1,

$$<u>^t = e^{-\frac{(1+\beta)\tau}{2}} \left(\frac{\tau}{(1+\beta)} \right)$$

4. **CONCLUSIONS**

Reports in the literature about appearance of an "overshoot" in the transient temperature for a finite slab were analyzed for the case of convective boundary condition with the wave term, . The solution from Fourier parabolic model predicts a monotonic rise in temperature with an asymptote of $<u>^{ss} = 1$ at steady state. The solution from the hyperbolic damped wave conduction and relaxation for average temperature of the finite slab was solved for by method of Laplace transforms for the time conditions such as the first derivative of temperature in time equal to zero at time zero as used by other investigators in similar problems with constant wall temperature boundary condition. The Laplace domain expression compares from this model compares well with the expression for prototypical second order systems as discussed in a process dynamics and control textbooks. This expression when inverted has a overshoot.

The temperature in a finite slab subject to convective boundary condition was assumed to comprise of a transient component and steady state component. The steady state component temperature was obtained and given by Eq. (52). The governing equation for transient temperature is rendered homogeneous upon obtaining the solution of steady state temperature. The transient temperature was expressed as a product of decaying exponential and wave temperature. The governing equation for the wave temperature was solved for. The integration constants were obtained from the initial temperature condition of $<u>^{t} = 0$ at zero time. The second integration constant was calculated by verifying that the model solution obeys the energy balance equation for average transient temperature. This leads to a solution for transient temperature that has a decaying exponential term and a cosinuous term with a phase lag, ϕ. Eq. (62) does not result in any overshoot as shown in Figures 3.0 – 5.0 for various values of frequency and ratio of thermal relaxation times between air and the solid. It was found that the dimensionless *Sharma Number* (heat) is given by $S^* = \left(\frac{h}{Sa}\right)$ is a significant parameter in the analysis. Expression for time taken to steady state was found and given by Eq. (63). No overshoot was found when the wave term was taken into account even for small values of β, the ratio of the relaxation time of air with that of the solid. This is an indication that the fundamental phenomena such as "acceleration of free electron" as discussed before [12] can be used to model transient heat conduction at short times and be within the second law of thermodynamics. The maxima in the transient temperature were found to *increase* with decreasing S* starting with large values such as 10. A cross-over was found after S* became less than about 2.2. Then the maxima in the average transient temperature was found to decrease with decrease in storage number, S*. Eq. (62) can be seen to go to zero in the asymptotic limits of infinite storage number, $S^* \rightarrow \infty$.

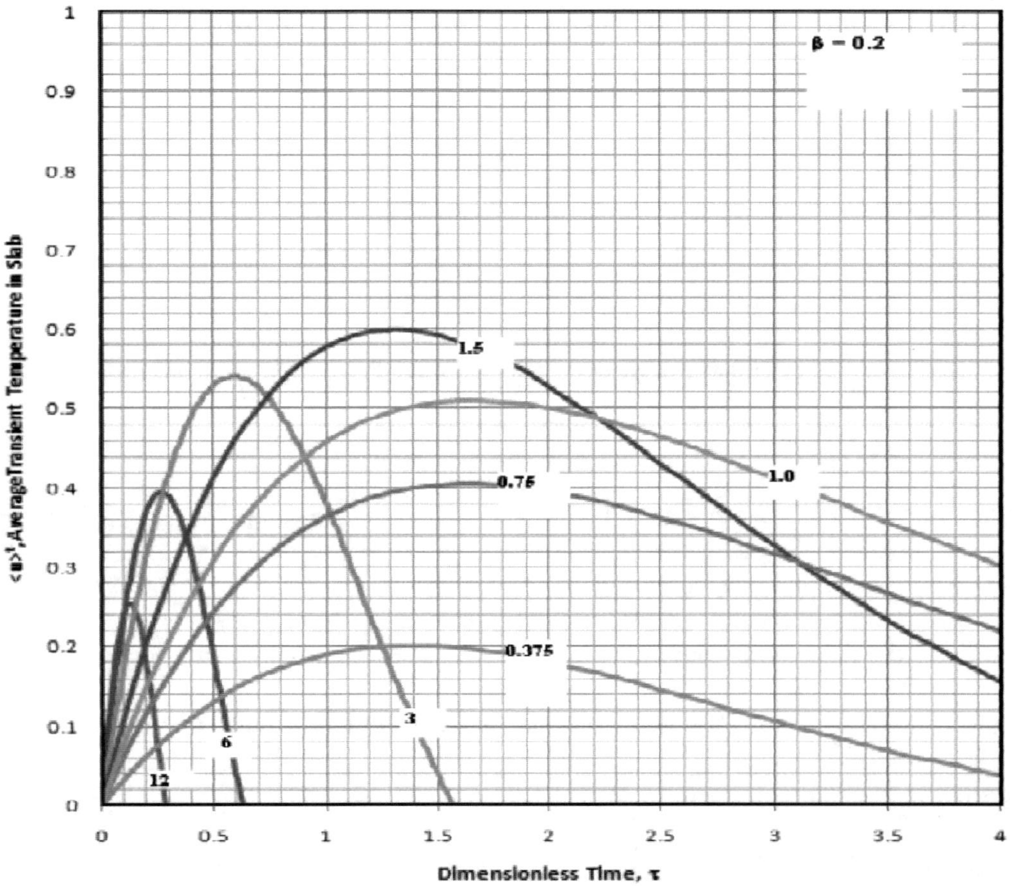

Figure 6.0 Transient Average Temperature in Finite Slab from Damped Wave Conduction and Relaxation (β =0.2)

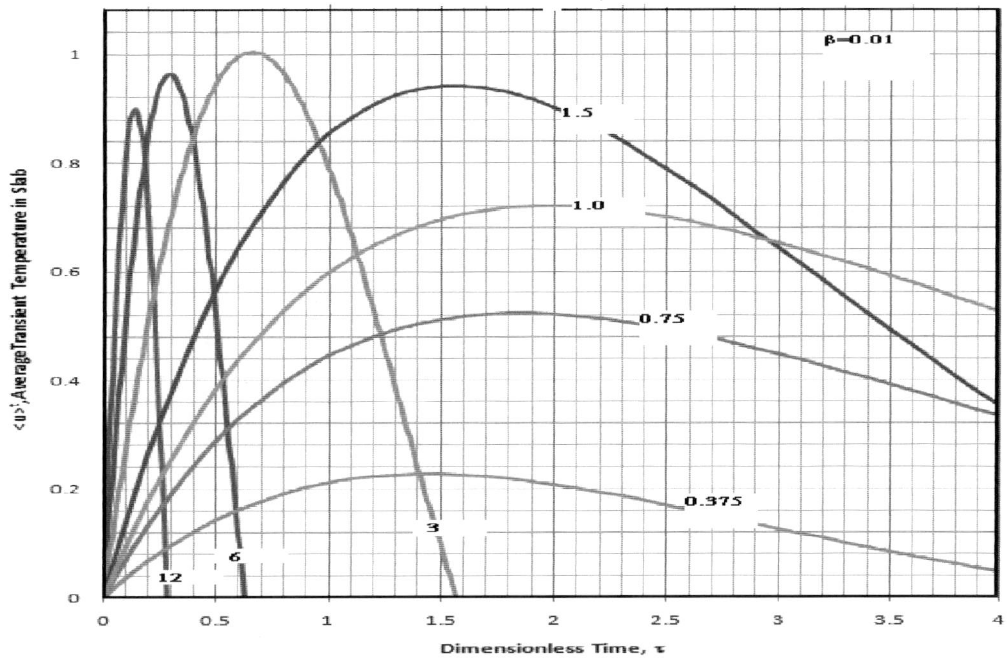

Figure 7.0 Transient Average Temperature in the Finite Slab Subject to Convective Boundary Condition with Derivative of Heat Flux with Time (β = 0.1)

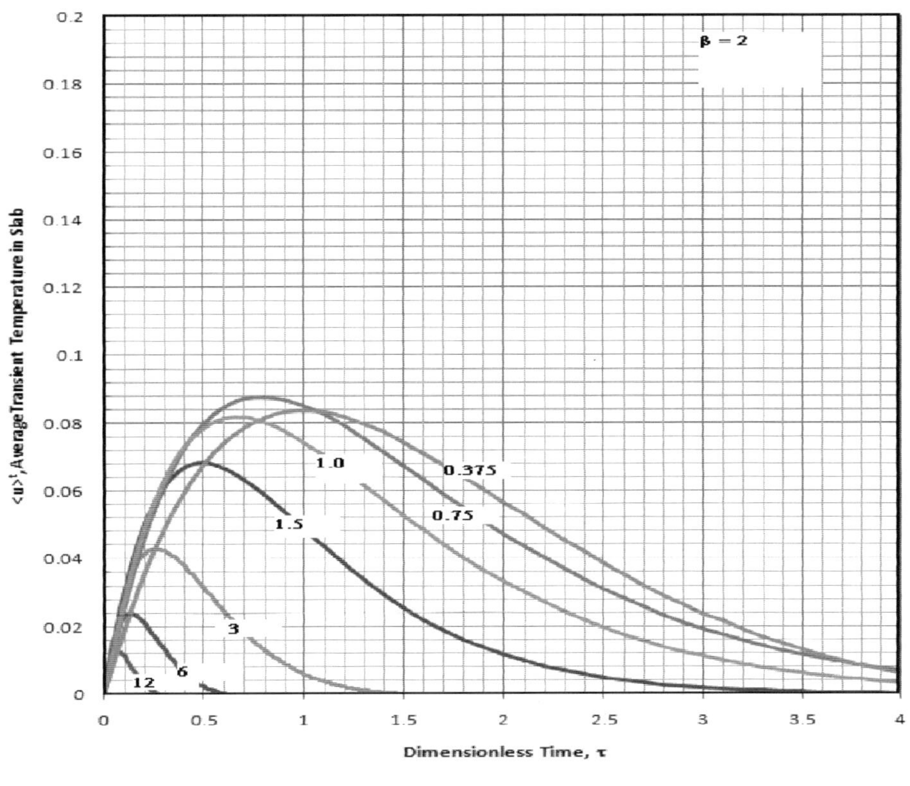

Figure 8.0 Transient Average Temperature in the Finite Slab Subject to Convective Boundary Condition with Derivative of Heat Flux with Time (β = 0.5)

5. Nomenclature

a	half-width of finite slab (m)
C_p	heat capacity ($J.Kg^{-1}.K^{-1}$)
g	function of X only
q	heat flux ($w.m^{-2}$)
T	temperarature (K)
T_s	surface temperature (K)
T_0	initial temperature (K)
V	function of time only
W	wave temperature, $u = We^{-\frac{\tau}{2}}$

Greek

α	thermal diffusivity ($m^2.s^{-1}$)
ρ	density ($kg\ m^{-3}$)
τ_r	relaxation time (s)
τ	dimensionless time (t/τ_r)

Dimensionless Groups

S* Sharma Number (heat), $S^* = \left(\frac{h}{Sa}\right)$

S Storage Number, $S = \left(\frac{\rho C_p}{\tau_r}\right)$

u Dimensionless temperature, $u = \left(\frac{T - T_s}{T_0 - T_s}\right)$

Ve Vernotte Number, $Ve = \left(\frac{\alpha \tau_r}{a^2}\right)$

X Dimensionless ordinate, $X = \left(\frac{x}{\sqrt{\alpha \tau_r}}\right)$

6. REFERENCES

[1] M. Al-Nimr, M. Naji, and S. Al-Wardat, "Overshooting Phenomenon in the Hyperbolic Heat Conduction Model", **Japanese Journal of Applied Physics, Part I: Regular Papers and Short Notes & Review Papers**, 42, 8, (2003) 5383-5386.

[2] M. A. Al-Nimr and M. Alkam, "Overshooting Phenomenon in the Hyperbolic Microscopic Heat Conduction Model", **Int. J. Thermophysics**, 24 (2), (2003), 577-583.

[3] Y. Taitel, On the Parabolic, Hyperbolic and Discrete Formulation of Heat Conduction Equation", **International Journal Of Heat and Mass Transfer**, Vol. 15, (2), (1972), 369-371.

[4] K. R. Sharma, On the Solution of Damped Wave Conduction and Relaxation Equation in a Semi-infinite Medium subject to Constant Wall Flux, **Int. Journal of Heat and Mass Transfer**, 51, (2008), 6024-6031.

[5] L. Onsager, Reciprocal Relations in Irreversible Processes, **Phys. Review**, 37, (1931), 405-426.

[6] K. R. Sharma, Manifestation of Acceleration during Transient Heat Conduction, **Journal of Thermophysics and Heat Transfer**, **26**, (4), (2006), 799-808.

[7] K. Renganathan, *Correlation of Heat Transfer with Pressure Fluctuations in Gas-Solid Fluidized Beds*, **Ph.D. Dissertation**, West Virginia University, Morgantown, WV, 1990.

[8] K. R. Sharma, Storage Coefficient of Substrate in a 2 GHz Microprocessor, *225th ACS National Meeting*, New Orleans, LA, March, 2003.

[9] K. R. Sharma, Third Order PDE in Restriction Fragment Length Gel Electrophoresis, *228th ACS National Meeting,* Philadelphia, PA, August 2004.

[10] K. R. Sharma, *Bioinformatics,* **McGraw Hill Professional**, (2009), New York, NY.

[11] K. R.Sharma, Adsorber Design for Removal of Acrylonitrile from Water, *AIChE Spring National Meeting,* New Orleans, LA, March /April, 2003.

[12] K. R. Sharma, Critical Radii Neither Greater than the Shape Limit Nor Less than Cycling Limit, *AIChE Spring National Meeting,* New Orleans, LA, March/April 2003. 23

[13] K. R. Sharma, Bessel Composite Function of the Third Order and First Kind: Solution to the Dissolving Pill Problem, *230th ACS National Meeting*, Washington, DC, August/September, 2005.

[14] H. B. G. Casimir, Note on the Conduction of Heat in Crystals, **Physica**, 5, **(**1938**),** 495 -500.

[15] W. Nernst, W., *Die Theoretician Grundalgen des n Warmestatzes*, Kanppe Hall, Frankfurt, 1917.

[16] T. Q. Qiu, C. L. Tien, Heat Transfer Mechanisms during Short Pulse Laser Heating of Metals, **ASME Transactions, Heat Transfer Division**, Vol. 196, (1992), 41-49.

[17] Y. S. Xie, Y. X. Yuan, X. B. Zhang, Hyperbolic Heat Conduction Equation and Analytic Solution for Instantaneous Ignition of Solid Propellant, **Bimggmg Xuebao/Acta Aramamentarii**, Vol. 27, Issue Supplement, March., (2006), 24-28.

[18] L. Landau, and E. M. Liftshitz, *Fluid Mechanics*, **Pergamon**, UK, 1987.

[19] J. C. Maxwell, On the Dynamical Theory of Gases, **Phil. Trans. Roy. Soc.**, 157, (1867), 49.

[20] C. Cattaneo, C., A Form of Heat Conduction which Eliminates the Paradox of Instantaneous Propagation, **Comptes Rendus**, **247**, (1958) 431- 433.

[21] C. Cattaneo, Sulla Coduzione del Calone, Atti. Sem. Fis. Univ. **Moderna,** 3 (1948), 83.

[22] P. Vernotte, Les Paradoxes de la Theorie Continue de l'equation de la Chaleur", **C. R.. Hebd. Seanc. Acad. Sci. Paris**, 246, (22), (1958), 3154 - 3155.

[23] D. D. Joseph, L. Preziosi, Heat Waves, **Reviews of Modern Physics**, 61, (1989), 41-73.

[24] D. D. Joseph, L. Preziosi, L., Addendum to Heat Waves, **Reviews of Modern Physics**, 62, (1990), 375-391.

[25] M. N. Ozisik, D. Y. Tzou, On the Wave Theory of Heat Conduction**, ASME J of Heat Transfer**, 116, (1994), 526- 535.

[26] D. Y. Tzou, *Macro to Microscale Heat Transfer: The Lagging Behavior*, **CRC Press**, New York, 1996.

[27] K. R. Sharma, *Damped Wave Transport and Relaxation*, **Elsevier**, Amsterdam, 2005.

[28] K. Mitra, S. Kumar, A. Vedavarz, M. K. Moallemi, Experimental Evidence of Hyperbolic Heat Conduction in Processed Meat, **Journal of Heat Transfer**, Vol. 117, (1995), 568-573.

[29] C. Bai, A. S. Lavine, On Hyperbolic Heat Conduction and Second Law of Thermodynamics, **Journal of Heat Transfer**, Vol. 117, (2), (1995), 256-263.

[30] E. Zanchini, Hyperbolic Heat Conduction Theories and non-Decreasing Entropy**, Phy. Review B-Condensed Matter and Material Physics**, Vol. 60, (2), (1999), 991-997.

[31] A. Barletta, E. Zanchini, Thermal-Wave Heat Conduction in a Solid Cylinder which undergoes a change of Boundary Temperature, **Heat and Mass Transfer/ Warema-und Stoffuebertragung**, 32, 4, (1997), 285-291. 24.

[32] M. A. Al-Nimr and O. M. Haddad, "The Dual-Phase-Lag Heat Conduction Model in Thin Slab under Fluctuating Thermal Disturbance", **Heat Transfer Engineering,** 24, 5, (2003), 47-54.

[33] M. A. Al-Nimr and M. Naji, "The Hyperbolic Heat Conduction Equation in an Anisotropic Material", **Int. J of Thermophysics**, 21, (2000), 281-287.

[34] A. M. Haji-Sheik, W. J. Minkowycz, and E. M. Sparrow, **J. Heat Transfer**, 124, 2, 307-319 (2002).

[35] P. J. Antaki, Solution for non-Fourier Dual Phase Lag Heat Conduction in a Semi-infinite Slab with Surface Heat Flux, **International Journal of Heat and Mass Transfer**, Vol. 41, (14), (1998), 2253-2258.

[36] K. R. Sharma, Temperature Solution in Semi-Infinite Medium under CWT using Cattaneo and Vernotte for non-Fourier Heat Conduction", *225th ACS National Meeting*, New Orleans, LA, March 23rd - March 28th, 2003.

[37] K. R. Sharma, Finite Speed Heat Conduction - III Relativistic Transformation from Cylindrical Coordinates in Semi- Infinite Medium, *Chemistry Preprint Archive*, Volume 2003, Issue 2, February 2003, 360-377.

[38] K. R. Sharma, Finite Speed Heat Conduction - IV Relativistic Transformation from Spherical Coordinates, *Chemistry Preprint Archive*, Volume 2003, Issue 2, February 2003, Pages 378-395.

[39] K. R. Sharma, A Fourth Mode of Heat Transfer called Damped Wave Conduction, *42nd Annual Convention of Chemists Meeting*, Santiniketan, India, February , 2006.

[40] K. R. Sharma, R. Turton, Mesoscopic Approach to Correlate Surface Heat Transfer Coefficients with Pressure Fluctuations in Gas-Solid Fluidized Beds, **Powder Technology**, Vol. 99, (2), (1998), 109-118.

[41] K. R. Sharma, On the Derivation of an Expression for Relaxation Time from Stokes-Einstein Relation, 233rd ACS National Meeting, Chicago, IL, March 2007, American Chemical Society, Washington, DC.

[42] B. W. Bequette, *Process Control: Modeling, Design and Simulation,* **Prentice Hall** (2003), Upper Saddle River, NJ.

[43] F. Golnaraghi, B. C. Kuo, *Automatic Control Systmes,* **John Wiley & Sons,** (2010), New York, NY.

(109b) Reactive Distillation Using External Side Reactors for Synthesis of Bio-Renewable Organic Acid Esters

Aspi K. Kolah[1], Carl T. Lira[1], Dennis J. Miller[1], John C. Prindle[2] and Chandrakant B. Panchal[2], (1)Chemical Engineering and Materials Science, Michigan State University, East Lansing, MI, (2)E3Tec Service LLC, Clarksville, MD

Organic acid esters, produced from bio-based organic acids and alcohols, are truly bio-renewable, "green" chemicals that are promising replacements for halogenated and other petroleum-based solvents, lubricants, and plasticizers. With current uncertainty regarding the future of ethanol and other alcohols as biofuels, and with the availability of organic acids via fermentation, there exists strong motivation to develop esters as viable commodity bio-chemicals to support and expand the existing bio-renewable infrastructure.

Esters are advantageously produced using reactive distillation, where simultaneous reaction and separation take place in the same process unit. The hardware design of standard reactive distillation columns exhibits challenges for slow chemical reactions where the reaction requirements of high liquid holdup and high catalyst mass must be present inside the reactive distillation column. A similar situation arises when desirable reaction temperature inside the reactive distillation column is not in consonance with the distillation temperature or when side reactions occur.

In order to make the synthesis of esters economically competitive with their petroleum-based counterparts, in this study we have evaluated an alternative to the classical reactive distillation concept by using a conventional distillation column integrated with external side reactors.

In this SBIR Phase I Project, we have developed and demonstrated advanced reactive distillation concepts of heat-integration with side reactors for esterification of citric acid with ethyl alcohol to produce tri-ethyl citrate (TEC), which is a good candidate for replacing phthalate-based plasticizers. Phthalates have potential carcinogeneity and overall health concerns to the extent that they have been banned in Europe from use in children's toys and human contact applications. The potential of TEC is currently limited by the lack of large-scale, energy efficient and economic production.

AspenPlus process analysis is performed based on the available kinetic data for esterification of citric acid with ethanol for an integration of pre-reactor as well as side reactors with fractionation column. Heat integration is examined for energy efficiency and maintaining temperatures in side reactors for maximizing the catalyst utilization. The concept of integrated reaction and distillation is validated by experiments using the reactive distillation facility at Michigan State University with a pre-reactor and one side reactor coupled to a distillation column.

Pilot Plant Testing – The Key to Solving Liquid-Liquid Extraction Challenges

Authors:

Donald J. Glatz
Thomas D. Lightfoot

Koch Modular Process Systems, LLC.

Prepared for Presentation at

AIChE Spring Meeting

San Antonio, TX
April – May 2013

INTRODUCTION

The complexity of the process taking place in a liquid-liquid extraction (LLE) column (e.g. droplet breakup and coalescing, mass transfer, and axial & radial mixing), and the fact that small amounts of impurities can dramatically affect column performance, make it difficult to design a production extraction column based upon computer models and simulations alone [1]. It remains problematic today to design LLE equipment with any reasonable degree of confidence and accuracy without some type of pilot plant testing. This paper will focus on how pilot plant testing (1) identified potential scale up challenges, and (2) how the testing led to solutions. The scale-up designs could only be guaranteed using the solutions found during the pilot plant testing. Based upon many years of experience with pilot testing LLE processes, specific challenges/solutions will be presented including unfavorable density gradients, changing hydraulic behavior throughout the column, negative heat of mixing, and transitional phase in the process. The fact that most of these items could only be identified and solved during testing will clearly support the importance of pilot plant testing new LLE processes.

APPLICATION 1 – Unfavorable Density Gradient

An unfavorable density gradient results when the density of the continuous phase within an extraction column is significantly higher at the top of the column than at the bottom of the column. This phenomenon will result in greater tendency for back-mixing and loss of efficiency as the column diameter increases. This is particularly problematic when attempting to scale up from a small diameter column in the pilot plant to a large diameter production column.

A client tested a 25 mm diameter KARR® Column for a process in which the aqueous feed contained >60% solute (product to be extracted) and a maximum amount of organic solvent was specified in order to make the process economical. Initial tests in the KARR® Column were successful, however very unfavorable density gradients were observed. From bottom to top of the column the specific gravity of the light phase changed from 0.867 to 0.924 and the specific gravity of the heavy phase changed from 1.01 to 1.04. In our experience the KARR® Column is not the optimal design for these types of processes due to the large open area and reciprocating action. Stationary baffling that restricts back-flow is better and as such we recommended the SCHEIBEL® Column, for additional testing and scale-up.

A summary of the optimal test results for this process in a 3" diameter x 60-stage SCHEIBEL® Column are shown here:

Run No.	Capacity GPH/ft^2	S/F Ratio Mass Ratio	Total % Recovery
1	600	1.0	98.9
2	600	1.3	99.2
3	600	1.5	99.5
4	450	1.3	97.8
5	650	1.3	99.4

Some key points from these results that were utilized in scale up to the production column:

- Excellent extraction efficiency (>99% recovery) was achieved using 60 agitated stages, a S/F ratio of 1.3, and design capacity of 600-650 GPH/ft^2.
- When the column was operated at a lower capacity (450 GPH/ft^2), recovery dropped to below 98%, whereas at higher capacity (650 GPH/ft^2) the recovery remained >99%. This is expected for systems with unfavorable density gradient as higher velocities lead to better plug flow tendency, and thus less back mixing. As such the production column must be designed and operated at the higher capacity.
- For the production column design, a 20% safety factor was applied. Thus, the production column was built with a total of 72 agitated stages.

APPLICATION 2 – Changing Hydraulic Behavior throughout the Column

Very often, the hydraulic behavior is significantly different as an extraction progresses through a counter current extraction column. This can be caused by several factors including changing flux rates, changing capacity, changing density difference between the phases and changing interfacial tension. This phenomenon is evident when observing the interaction between the phases in a glass shell column. Compensation can be made by varying the mixing intensity throughout the column. For a column with rotating internals, this is often achieved by changing the diameter of the impellers. For a KARR® Column this is often achieved by varying the spacing between the perforated plates. Systems with a high amount of mass transfer (greater than 10% mass transfer), are obvious candidates for implementing variations prior to testing. For these types of systems, experience can lead to an "educated" guess as to the variation that will work. However, for some systems, it is not obvious that modified agitation intensity will be required.

An example of this was a system for extraction of methanol from an acrylate feed stream using water as the solvent. The organic feed contained approximately 2.5% methanol and the target raffinate concentration was less than 0.1% methanol. Liquid-liquid equilibrium data showed that approximately 6 theoretical stages would be required to achieve this result at the desired S/F of 0.3. For testing, a 25 mm KARR® Column was set up with 6' plate stack height. A summary of the results from testing in this column is shown here:

Run No.	Plate Stack	Feed Rate (ml/min)	Solvent Rate (ml/min)	Agitation SPM	Raffinate % MeOH
1	Standard plate spacing	150	45	75	0.165
2	Standard plate spacing	150	45	100	0.124
3	Standard plate spacing	150	45	110	Flooded
4	Variable plate spacing	180	54	200	0.078
5	Variable plate spacing	210	63	175	0.084

For standard plate spacing (2" spacing throughout the column), the minimum raffinate concentration was 0.124% at an agitation speed of 100 strokes per minute (SPM). Attempts to run at a higher agitation speed resulted in flooding. It was observed that the dispersed water phase droplets would fall very quickly through the first 4' of the column with very little hold-up. At >100 SPM, these droplets would start to back up and flood about 1-2 feet above the organic feed inlet. Thus, although there was only 2.5% mass transfer, the area immediately above the feed inlet was obviously the controlling region in the column. Prior to Run 4, the following plate stack was installed into the column; from the bottom, 1' of 6" spacing, 1' of 4" spacing and then 4' of the standard 2" spacing. As can be seen from the additional runs, this revised plate spacing allowed for both increased agitation speed and increased capacity. The higher agitation speed was required in order to achieve the desired < 0.1% methanol in the raffinate.

Several production columns have been designed from this pilot test work, and these are operating successfully today.

APPLICATION 3 – Negative Heat of Mixing

A process was tested in which an organic salt solution was fed to the bottom of an extraction column and chloroform was added at the top. Based upon the process, the client could control the feed such that the amount of organic product was in the range of 20-45 wt% and the salt content was in the range of 10-30 wt%.

The first feed tested was 40% organic product and 15% salt. The column was a 3" diameter SCHEIBEL® Column and the target operating temperature was 40 °C. For many runs, this column gave excellent extraction results, and based upon the hydraulic behavior inside the column, it was the best column design for this process. As testing progressed, capacity was pushed and lower solvent to feed ratios were attempted. Eventually, a cold spot became evident in the middle of the column (approximately 20 °C). This cold spot resulted in salt dropping out of solution and caused the column to flood. It was determined, that the transfer of the organic product from the aqueous phase to the chloroform phase was accomplished with a negative heat-of-solution, causing the temperature reduction in the middle of the column.

An experimental design was set up to evaluate the following variables; organic product concentration in the feed, salt concentration and solvent to feed ratio. Based upon the results from this testing, an allowable operating window was determined. Included in this window were the conditions that would produce the desired extraction result defined as the maximum allowable amount of product in the raffinate stream (lost product). From an extraction efficiency standpoint, lower solvent to feed ratios could produce the desired product recovery. However, minimum ratios were identified (based upon the feed composition) in order to prevent the "salting out" phenomenon and subsequent flooding.

APPLICATION 4 – Transition Phase in the Process

Extraction processes for recovery of metals are often accomplished via a multi-step process in which the metal ion is first recovered from an acidic "leach" solution using an organic solvent mixture following by back-extraction or stripping the metal from the organic phase using an aqueous "strip" solution. In these processes, the metal ions are essentially removed from the impurities in the original leach solution, and then put back into the "clean" aqueous phase from which they can be utilized as highly pure materials.

In one particular process, a metal ion was extracted from the leach solution very successfully using a SCHEIBEL® Column. This column worked efficiently due to a high density difference and high interfacial tension, for which the high agitation and internal baffling of the SCHEIBEL® Column provided the drop size and hold-up necessary to achieve the metal ion extraction. The stripping step was another matter. Upon start up for the stripping step, the concentrations of components in the process were such that side reactions occurred, resulting in a large amount of by-product formation. Insoluble solids were generated, having an appearance of white and fluffy material. Attempts to test this in columns with rotating internals resulted in plugging and flooding. Even mixer-settlers proved to be problematic for this process. Although not tested, it was determined that a packed or trayed column would not be able to handle the transition phase due to plugging.

Next the KARR® Column was tested. The benefits of the KARR® Column for this type of process were that the plate stack design is very open (55-60% open area) and the fact that the entire plate stack is reciprocating and thus there are no "dead" zones where solids can build up. Upon start up in the KARR® Column, the transition phase was once again observed by fine solids dropping out of solution. As hoped, these solids did not find any place to hold up in the column and flowed out of the column with the aqueous phase. After several minutes of operation, the system moved past the transition phase and no more solids were observed to form in the process. As the system reached steady state operation in the column, the entire column cleared up and clear solutions were observed to exit at both ends of the column. The extraction efficiency was excellent (complete stripping accomplished) and the KARR® Column was obviously the best choice for this process step.

APPLICATION 5 – Emulsion Formation

Emulsion formation in an extraction process can be a difficult problem for testing and design. The KARR® Column been demonstrated to be an excellent choice for these types of systems including fermentation and algae broths. A previous AIChE paper provides a good reference for this [3].

APPLICATION 6 – Agitation vs No Agitation

In some applications, agitation has been required in order to allow the two phases to pass counter currently through an extraction column, whereas flooding and interface control problems occur without agitation. This phenomenon was also covered in a paper presented at a previous AIChE meeting [4].

CONCLUSION

It is clear from the observations and results presented above, that many potential problems that can be encountered in extraction processes can only be identified during pilot plant testing. As shown in the examples presented, it was the observations and results during pilot plant testing that allowed for the accurate design of the production columns.

LITERATURE CITED

1. Cusack, R., Glatz, D., "Apply Liquid-Liquid Extraction to Today's Problems", Chemical Engineering, July 1996.

2. Cusack, R., Fremeaux, P., and Glatz, D., "A Fresh Look at Liquid-Liquid Extraction, Extractor Design and Specification", Chemical Engineering, February 1991.

3. Glatz, D., and Lightfoot, T., "Using Extraction Columns for Recovery of Valuable Products from Biomass", AIChE Fall National Meeting, October – November 2012.

4. Glatz, D., and Mason, L., "Extraction of Methanol and Glycerin from Crude Biodiesel", AIChE Fall National Meeting, November 5-9, 2007.

(109d) Development of a Standardized Method for Mass Transfer Measurements in Absorption and Desorption

Anna-Katharina Kunze[1], Philip Lutze[1], Manuela Kopatschek[2], Stephan Mueller[2], Marcus Gruenewald[2] and Andrzej Górak[3], (1)Department of Biochemical and Chemical Engineering, Laboratory of Fluid Separations, TU Dortmund University, D-44227 Dortmund, Germany, (2)Mechanical Engineering Department, University of Bochum, Bochum, Germany, (3)Department of Biochemical and Chemical Engineering, Laboratory of Fluid Separations, TU Dortmund University, Dortmund, Germany

Development of a standardized method for mass transfer measurements in absorption and desorption

A. Kunze[1], P. Lutze[1], M. Kopatschek[2], S. Müller[2], M. Grünewald[2], A. Górak[1]

[1] TU Dortmund University, Laboratory of Fluid Separations

Emil-Figge-Strasse 70, D-44227 Dortmund, Germany

Phone: +49 (0) 231 755-3034, E-Mail: anna-katharina.kunze@bci.tu-dortmund.de

[2] University of Bochum, Chair of Fluid Separations

Universitaetsstrasse 150, D-44780 Bochum, Germany

Due to the rising use of rigorous models for designing absorption and desorption processes a coherent method for mass transfer parameter determination must be developed. Currently, a comparability of mass transfer parameters investigated experimentally at different sights or different columns is not existent. Additionally, there is hardly one method to evaluate mass transfer parameters determined in the past, because often important information about the measurements and calculation is missing. That leads to the effort of a standardized experimental and calculation method for determination of mass transfer parameters in absorption and desorption. The main focus is to increase transparency and comparability for mass transfer parameters on gas and liquid side $\boldsymbol{\beta_G}$ und $\boldsymbol{\beta_L}$ and the effective interfacial area $\boldsymbol{a_{eff}}$ from different authors[1,2]. This will lead to a decrease in necessary safety factors resulting in a cost and resource optimized column design.

A consortium of 14 industrial partners, including manufacturers and users of packings, as well as two academic partners are working on this standard method in Europe. The first step of the standard method is to figure out an experimental method to achieve coherent results on pilot plant scale. Therefore, concentration measurements on different plants are carried out to calculate the volumetric mass transfer parameters $\boldsymbol{(\beta_G \cdot a_{eff})}$ and $\boldsymbol{(\beta_L \cdot a_{eff})}$. The method is developed for structured and random packings. During the development of an experimental method the main focus is on the configuration of the setup. Detailed investigations concerning the influence of e.g. gas and liquid distributors, packing height, and taking samples is carried out. The experiments for the experimental method are carried out in two columns with a diameter of 288 mm, which have exactly the same configuration. The experimental method will then be adapted additionally to a column with a higher diameter to carry out further experiments.

In addition a mathematical tool is necessary, which guarantees a consistent calculation of mass transfer parameters for rigorous modeling. The stepwise data processing from the experimental data to gain parameters for rigorous modeling is described in Figure 1.

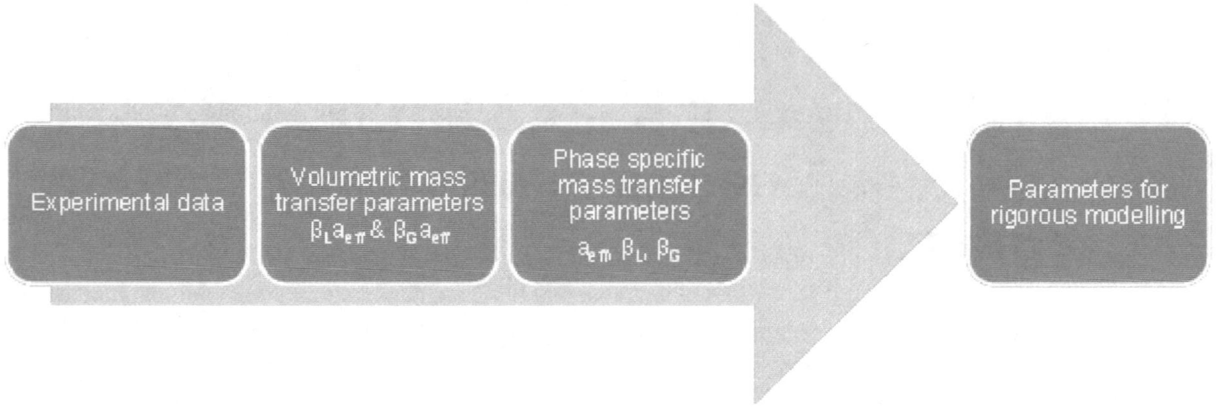

Figure 1: Pathway for the determination of model parameters of rigorous absorption models.

Based on the experimental data the first step is to define the volumetric mass transfer parameters **($\beta_G \cdot a_{eff}$)** and **($\beta_L \cdot a_{eff}$)**. To take temperature and concentration dependence of the chemical system data into account, a detailed analysis of possible calculation methods is carried out. To define the temperature and concentration dependence of the phase equilibrium, at least seven basic system data (such as vapor pressure as well as density, viscosity and diffusion coefficients in gas and liquid phase) have to be calculated. To minimize the calculation effort, a global sensitivity analysis is carried out. For the most sensitive parameters, different calculation methods are compared. In addition, also the experimental data are implemented in the sensitivity analysis to determine the influence of deviations in each measured value.

A combination of three chemical systems is necessary to determine β_L, β_G and a_{eff}, one of the following properties has to be fulfilled by one of the systems:

- mass transfer resistance on liquid side for β_L
- mass transfer resistance on gas side for β_G
- fast reaction pseudo first order for interfacial area a_{eff}

Via correlation of interfacial area between the three different chemical systems the phase specific mass transfer parameters can be determined.

Within this presentation, the detailed results of the analysis of equipment parameters, the method in general as well as the mathematical tool will be shown.

This work is part of GVT funded project "Standardization of mass transfer measurements in absorption and desorption".

Literature

[1] A. Hoffmann et al., "Standardization of mass transfer measurements: A basis for the description of absorption processes", Chemical Engineering Research and Design 2007, 85 (1), 40-49.

[2] J.F. Rejl et al., "Methods standardization in the measurement of mass-transfer characteristics in packed absorption columns", Chemical Engineering Research and Design 2009, 87 (5), 695-704.

A REVIEW OF THE ROTARY PRESSURE FILTER FOR THE FILTRATION OF A LIQUEFIED GAS SLURRY

Barry A. Perlmutter
BHS-Filtration Inc.
Charlotte, North Carolina 28270

Biography
Barry A. Perlmutter is President and Managing Director of BHS-Filtration Inc., a subsidiary of BHS-Sonthofen GmbH. Barry has over 30 years of technical and marketing experience in the field of solid-liquid separation, filtration, centrifugation and drying. He has published and lectured worldwide and has been responsible for introducing and creating growth for many European companies into the marketplace. He has a BS in Chemistry from Albany State, NY, MS from the School of Engineering at Washington University, St. Louis and an MBA from the University of Illinois.

ABSTRACT

In the manufacturing of this specialty chemical, the pilot plant examined the process to determine the most effective method for filtration to develop a production process. A study was undertaken, by plant engineering and development, to examine the use of a rotary pressure filter to develop the overall process operations for product quality and production rates.

This paper reviews the basics of two alternatives: rotary pressure filter and pressurized drum filters. The technologies are reviewed for the filtration of a liquefied gas slurry.

The article continues with the discussion of the process testing in the laboratory and in the field to evaluate the Rotary Pressure Filter. The lab testing used a BHS pressurized pocket leaf filter with 20 cm^2 of filter area while the fieldwork was conducted on a Rotary Pressure Filter with a filter area of 0.18 m^2.

The paper discusses test procedures, data collection, filter media selection and other process parameters. Scale-up data includes process conditions and production rates.

The paper conclusion is two-fold: a review of the unique specifications of the rotary pressure filter required for the process to meet production rates, product quality and site environmental and safety concerns. Finally, the paper demonstrates the creative process solution that is developed with close collaboration between the operating company and the vendor.

Acknowledgements: The author would like to thank Brian Moore, BHS Application Sales Engineer and Bruce Glines, BHS Senior Process Engineer for their help and assistance with this article.

LIQUEFIED GAS SLURRY

A liquefied gas is a clean, colorless gas that is easy to liquefy and transport. It is gaseous at normal temperature and pressure, but changes to a liquid when subjected to modest pressure or cooling. This easy liquefaction makes it easy to transport and store. This and other properties of certain liquefied gases, including a high oxygen content, lack of sulfur or other noxious compounds, and ultra clean make these versatile and promising process solvents. The most important point is to keep these gases, under pressure, so they behave as a liquid during filtration, cake washing and drying. The plant engineers reviewed the Rotary Pressure Filter and the Pressurized Drum Filter as two possible alternative solutions.

ROTARY PRESSURE FILTER (RPF)

The rotary pressure filter, shown in Figure 1, is a continuous pressure filter designed for thin cake to deep cake filtration with cake depths from 6 mm to 150 mm. The slowly rotating drum (6 – 60 rph), in Figure 2, is divided into segments (called cells) each with their own filter media (synthetic cloth or single or multilayer metal) and outlet for filtrate or gas. The outlets are manifolded internally to a service/control head where each stream can be directed to a specific plant piping scheme or collection tank. In this way, the mother liquor can be kept separate from the subsequent washing filtrates and drying gases. This allows for better process control as well as reuse and recovery of solvents and the gases. The service/control head, for this application, is pressure rated so the liquefied gas can be kept under pressure, acting as a liquid.

Figure 3 shows a typical operation as follows. The feed suspension enters each cell, under constant pressure, to form a filter cake. Internal divisions of the housing allow the cake to be processed in completely separate zones. Each zone can operate under different pressures depending upon the compressibility of the cake at each stage of filtration, washing and drying.

For example, pressure filtration is conducted up to a maximum differential pressure of 6 bar. Pressure filtration has the added benefit of eliminating post-precipitation of the solids if the process solvent is prone to flashing under vacuum filtration. In this case, the 6 bar pressure differential allows for a back-pressure of 2-bar or more, such that filtration can be at 4 bar and the gas can be a liquid.

A second benefit of the slow-rotating pressure filtration is the reproducible cake depth and the ability to control the residence time (by the speed of the drum). The slow rotation also results in much lower maintenance and energy costs.

Positive displacement washing or counter-current washing follows filtration. The design of the pressure cells, such that they have no free-space and are completely filled, eliminates the possibility of cake cracking or bypass, providing maximum washing efficiency. Multiple washing steps can be accomplished along with solvent exchanges, steaming and extraction.

The cake then undergoes a drying step by blowing ambient temperature or hot gas through the cake. The control of this drying step is influenced by gas flow, pressure, temperature as well as the cake depth and residence time.

Finally, the pre-dried cake is continuously discharged. The cake chute is pressurized which allows the cake to be fed directly into the downstream dryer without the use of rotary locks. This is an important benefit to keep the entire system pressurized. The cake discharge is assisted by a gas pulse.

After cake discharge, there is a cloth rinse step. This is controllable depending upon the product discharge, filter media and potential of cloth blinding. With the drum rotating, after the cloth rinse step, the cells are ready to be filled for filtration. A typical process flow is shown graphically in Figure 3.

PRESSURIZED DRUM FILTER (P-DF)

The other alternative to the RPF is the pressurized drum filter (P-DF) which is a rotating drum inside a pressure vessel. The unit consists of a filter drum, slurry trough, agitator, wash bars and a "pressure let-down rotary valve. The process begins by closing the pressure vessel manually with bolts and nuts, pressurizing the vessel with compressed gas, the rotary valve is also pressurized for sealing and the filter trough is filled via the suspension feed pipe and the agitator is started to keep the solids in suspension.

During the evaluation stage, the plant engineers reviewed the P-DF operation, as described above, and determined that the gas usage for the pressure vessel and rotary valve was very excessive. This is not required in the RPF.

Secondly, the RPF housing does not require the manual process of closing and sealing which saves man-hours and time as well as eliminates a possible leak path of the liquefied gas.

Finally, in the RPF, the solids fill the cell via a centrifugal feed pump which eliminates the need for a filter trough, agitator and feed pipe. It was for these points as well as the additional safety relief valves required, the back pressure to maintain the gas as a liquid and other points that the decision was to begin the test work for the RPF.

ON-SITE PROCESS TESTWORK: POCKET LEAF FILTER (PLF) TESTWORK

Process testing was conducted at the site's laboratory and in the plant. For the bench-top lab testing, the BHS pressurized pocket-leaf filter (PLF) with 20 cm^2 of filter area is used as shown in Figure 4. For the continuous pilot testing, a pilot RPF with 0.18 m^2 of filter area is installed, as shown in Figure 5.

The objectives of the PLF testing are as follows:

Filtration time vs. filter media
Filtration time vs. slurry feed mass
Filtration time vs. differential pressure
Filtrate quality vs. filter media
Cake solids wash time and quality
Cake solids discharge characteristics
Production Scale-Up and Process Guarantee

The lab testing proved to be uniquely challenging both to feed the PLF as well as to maintain a pressure to keep the gas as a liquefied solvent. The plant engineers and BHS developed a confidential method to meet these challenges.

CONCLUSION AND PATH FORWARD STEPS FROM THE PLF TESTING

The PLF tests demonstrated that acceptable filtration and solids wash rates could be obtained for this product and acceptable solids levels were observed for the mother liquor filtrate. Washing results and drying quality parameters were also achieved.

Additional pilot plant tests with the BHS continuous pilot unit, RPF 0.18, are recommended to confirm the PLF lab tests. In these tests, BHS would be able to identify the necessary slurry solids percentage, cake solids thickness, solids wash time, solids drying time as well as cake discharge. Finally, the pilot testing will be the basis for the mechanical design of the RPF to ensure that the RPF can be designed for the process with a liquefied gas slurry.

DISCUSSION OF THE RPF 0.18 M^2 PILOT TESTWORK AND SCALE-UP

While the actual data is confidential, the plant engineers and BHS process engineers gathered the following parameters:

Process Parameters:

- Slurry Feed Pressure:
- Slurry Feed Flow:
- Wash Pressure:
- Wash Flow:
- Dry Pressure:
- Drying Air Flow:

RPF Parameters:

- Drum Speed:
- Separating elements:
- Cake blow back:
- Cloth blow back:
- Backpressure:
- Cake Thickness:
- Filter Cloth:

To fully evaluate the RPF performance, the site also compiled the following:

- Slurry solids concentration
- Filtrate quantity (mother liquor, wash, blow down, etc.)
- Filtrate yield
- Cake Moisture
- Total Cake quantity

Scale-Up From RPF 0.18 M^2 Pilot Data

- Calculate Specific Filter Performance from Pilot Testing
  = kg of dry solids/m^2/hour
- Calculate Production Area Required from Filter Performance and Client Required Production Rate
- Using the drum speed, time for filtration, washing and drying and several other RPF factors, the specific filter area is calculated

TYPICAL SCALE-UP CALCULATION - EXAMPLE ONLY

The scale-up is based on 224 g slurry with 1:1 composition = 112 g dry solids:

Drum revolutions:

$$n = \frac{3600\frac{s}{h}}{(4+8+15)s} \cdot \frac{270°}{360°} \cdot 0,85 = 85\frac{\text{revolutions}}{h}$$

270°: active angle
0.85: factor for separating elements

Specific filter performance:

$$Q = 500\frac{1}{m^2} \cdot 85\frac{1}{h} \cdot 112g = 4760\frac{kg\ dry\ solids}{m^2 \cdot h}$$

Required filter area:

$$A = \frac{20000\frac{kgdrysolids}{h}}{4760\frac{kgdrysolids}{m^2h}} = 4,2\ m^2$$

Selected filter: BHS-Rotary Pressure Filter, type B16 with 5.4 m^2 is sufficient to operate 20,000 kg dry solids per hour.

DESCRIPTION AND SPECIFICATIONS OF THE RPF

The RPF design has several unique features for this application to ensure that the liquifeid gas remains as a liquid and secondly to ensure the safe operation of the unit.

- Direct-coupled drive for completely stable drum rotation, shown Figure 6.
- Closed and pressurized housing and bearing shields, shown in Figure 7.
- Secondary containment of the control head
- Pressurized control head
- Pressurized discharge housing

SUMMARY

As operating companies develop new and unique chemical processes, there are many choices for filtration, cake washing and drying. Engineers must evaluate all outcomes to make an informed and successful decision. Technical evaluation and, as shown, laboratory and pilot testing are critical for a successful decision and project. The take-away is that close collaboration between the operating company and the vendor will allow for creative problem-solving and process filtration solutions to achieve the desired quality and production requirements.

Figure 1: RPF Filter

FIGURE 2: RPF DRUM AND CELLS

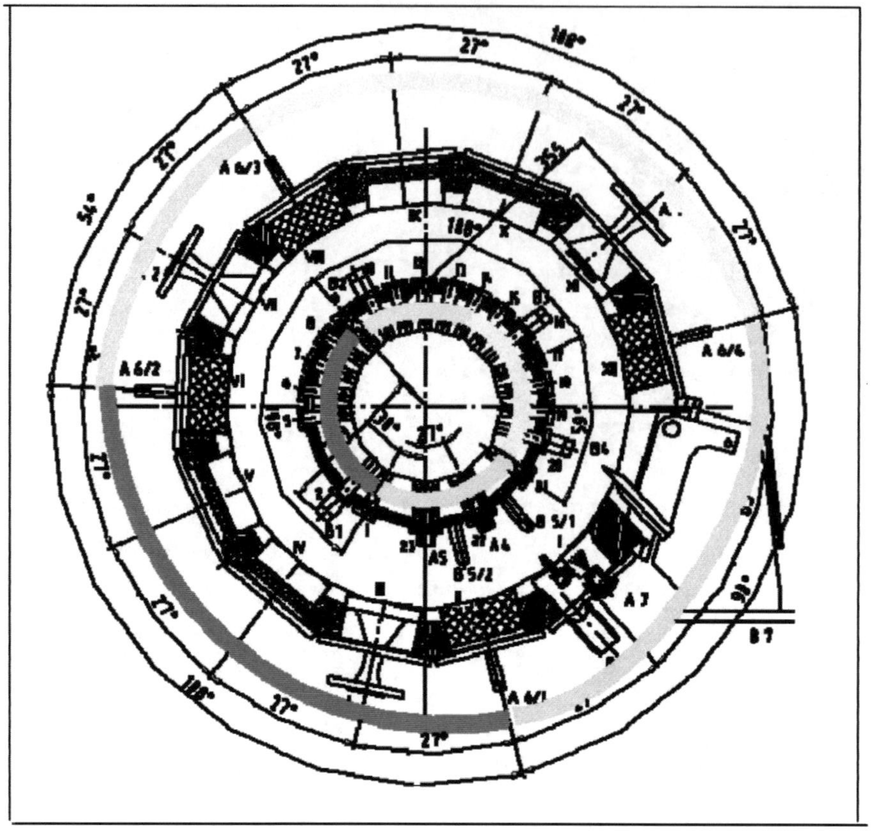

FIGURE 3: PROCESS FLOW

FILTRATION (RED)
CAKE WASHING (BLUE)
DRYING (YELLOW)
CAKE DISCHARGE & CLOTH RINSE (GREEN)

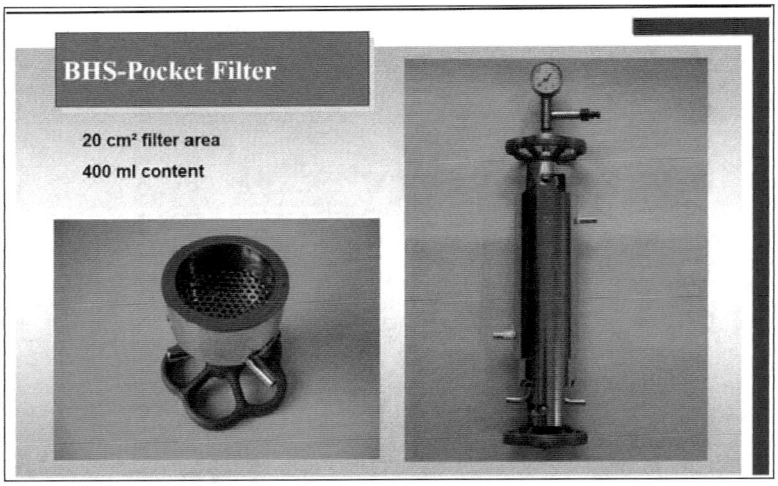

FIGURE 4: BHS POCKET LEAF FILTER (PLF)

FIGURE 5: RPF 0.18 M^2 PILOT FILTER

FIGURE 6: RPF DIRECT-COUPLED DRIVE

FIGURE 7: RPF PRESSURIZED HOUSING AND BEARING SHIELDS

(110a) Acid Gas Absorption Fundamentals

Clint P. Aichele[1], James R. Whiteley[1], Yash Tamhankar[1] and M.R. Resetarits[2], (1)School of Chemical Engineering, Oklahoma State University, Stillwater, OK, (2)Fractionation Research, Inc., Stillwater, OK

Absorption is a standard unit operation that has been studied by chemical engineering students for decades. One of its primary uses is for the removal of acid gases, like CO_2 and H_2S, from natural gas streams. Historically, aqueous amine and hot carbonate solutions have been used to perform these absorptions. Many different amines, solution strengths, activators and passivators have been tested and employed industrially. Approximately half of the industrial towers are trayed; half, packed. Foaming adversely affects the hydraulic performance in these systems.

Acid gas absorption is receiving appreciable new attention as concerns rise regarding CO_2 in the atmosphere. In addition, the United States has recently experienced significant growth opportunities in shale gas production which require improvements in gas treating technologies. This presentation reviews acid gas absorption and regeneration fundamentals. Several PFD's will be shown to illustrate important concepts. New absorbent solutions will be discussed, including ionic liquids. Flue gas applications will be discussed, including the difficulties associated with column pressure drops and very large column diameters. Rate-based modeling approaches will also be reviewed.

(110b) Sorption and Diffusion of Ethanol and Water in Fluoride Mediated Silicalite-1

Ke Zhang[1], Michelle Dose[1], Ryan P. Lively[2], William J. Koros[3] and Ron R Chance[4], (1)School of Chemical and Biomolecular Engineering, Georgia Institute of Technology, Atlanta, GA, (2)Algenol Biofuels, Bonita Springs, FL, (3)School of Chemical & Biomolecular Engineering, Georgia Institute of Technology, Atlanta, GA, (4)Chemical and Biomolecular Engineering, Georgia Tech, Atlanta, GA

Sorption and diffusion of ethanol and water have been studied in silicalite-1 crystals synthesized via a fluoride mediated route. The near-perfect hydrophobic silicalite (F^-) crystals have very few internal silanol defects and show unusually attractive ethanol and water transport behavior. Full isotherms (0.05 – 0.95 activity) over the range of 25- 55 ^oC are presented. At a water activity level of 0.95 (35 ^oC), the total water uptake by silicalite-1 (F^-) was found to be 0.263 mmol/g, a factor of 8 lower water uptake than in silicalite-1 (OH^-), thereby giving an ideal sorption selectivity of 35 for 5 mol% ethanol solutions. Uptake rates for water and ethanol have been measured in silicalite-1 (F^-) over a wide range of activities between 25-45 ^oC. The transient uptake curves are consistent with intra-crystalline diffusion control and yield transport diffusivities (D_t) of order 10^{-9}-10^{-10} cm^2s^{-1} for ethanol and water, with the values for ethanol being about three times larger than those for water at comparable low activity conditions. Thermodynamically corrected diffusivities (D_o) for ethanol, which reflect the intrinsic mobility, are almost an order of magnitude smaller than the values for water, as is expected from the difference in molecular size. Such results yield ethanol/water permselectivities of 7-27 for 1-5 mol% ethanol solutions and estimated ideal separation factors of 130-310 over the same concentration range. For binary ethanol-water mixture tests with competitive sorption, lower actual separation factors are expected, since the sorbed ethanol molecules may act as water sorption site "seeds" to decrease ethanol/water sorption selectivity. However it is clear that silicalite (F^-) is promising for ethanol extraction from water due to its significantly depressed water adsorption compared to silicalite (OH^-).

(110c) Distillation and Absorption Equipment

Mike Resetarits, Anand N. Vennavelli and A. Y. Ogundeji, Fractionation Research, Inc., Stillwater, OK

Distillation is almost as old as boiling, and yet, it remains the primary technique whereby chemicals are separated. Absorption is the primary technique whereby acid gases like CO2 are removed from gas streams; carbon capture is in the news. Distillation and absorption columns still dominate the skylines of refineries, chemical plants, and gas plants. The columns almost always contain trays or packings. Fractionation Research, Inc. (FRI) is a global consortium of companies that fund tray and packing research in Stillwater, OK. New generations of these devices have allowed column owners to revamp existing columns to achieve capacity increases, separation improvements, and energy consumption decreases. Augmented crossflow, counter-flow, and co-current flow trays have exhibited capacity increases. New generations of random packings and the associated distributors haven exhibited separation improvements. Structured packings and new high capacity versions have revolutionized vacuum towers where low-pressure drops are critical. This tutorial session reviews the newest trays and packings and FRI's testing of them

(110d) Radioisotope Methods for Obtaining Process Information

Lowell Pless, Tracerco, Pasadena, TX

In all chemical plants and refineries there generally is not enough instrumentation in the right places in our processes. During the design phase, decisions were made between knowing more about the process, which means buying more instruments, and saving capital funds. When a problem appears with the operation of a vessel, there are always gaps in the knowledge of what is happening due to a lack of instruments in the places where knowledge is not routinely required. Process engineers and operations personnel need to know what is happening so they can take corrective action. When knowledge is missing, they often take actions that make matters worse.

For distillation and separation processes Chemical Engineers have historically relied upon plant process measurements such as flow rates, temperatures, and pressures and model or simulation results for data with which to solve troubleshooting or re-design problems. An array of on-line diagnostic services are available that can provide additional data offering real-time information on how pieces of process equipment are actually operating.

The most common application is G***amma Scanning***. This test is primarily applied, but not necessarily limited to, distillation or separation columns. Gamma scans provide a density profile of the internal process of operating distillation columns and other process vessels. The density profile can be used to diagnose the hydraulic operating conditions of mass transfer devices such as damage to internals, flooding, degree of entrainment or weeping, liquid levels on trays and distributors, liquid distribution through packed beds, etc. The presenter will show case studies where scanning revealed vital process information that helped solve an operating problem, or helped make a revamp/re-design successful.

Additional derivatives from gamma scanning of detecting flow distribution patterns will be demonstrated. The ***CAT-Scan*** technique is a specialized horizontal gamma scan used to generate a topographic profile of the internal cross-sectional density of process equipment. This profile is useful for the detailed study of liquid flow distribution through packed columns as well as having applications beyond distillation columns. A hybrid application involving tracers and scanning on fixed bed catalytic reactors will also be discussed.

Another technique involving radioactive materials is the ***Tracer Test***, in which a radioactive tracer is introduced into the process for investigating the process flows. External radiation detectors are placed at strategically chosen locations outside vessels and piping to monitor the tracer flow through the process. Either liquid, gas or a solid tracer can be utilized depending on process conditions and the fluid to be studied. This tracer technique can identify liquid or gas maldistribution in fixed or fluidized bed reactors. The residence time through a tank or reactor can be measured and compared with the calculated residence time. Observed differences can relate directly to the degree of mixing or channeling inside the reactor. These tracer techniques have also been used in identifying leakage of heat exchangers and valves.

(110e) Liquid Extraction Processes & Equipment

Mike Resetarits, A. Y. Ogundeji and T.J. Cai, Fractionation Research, Inc., Stillwater, OK

What is the best way to remove C_6, C_7 and C_8 compounds from a feed stream containing C_6, C_7 and C_8compounds? How do you take a lubricating oil and remove certain compounds and improve its viscosity index (so that its viscosity does not change excessively with the oil's temperature)? What is propane deasphalting? The answers to all of these questions in "liquid extraction processes."

What technology was supported and funded by the actor Kevin Costner to help clean up the large 2010 Gulf of Mexico oil spill? What are Karr reciprocating plate columns, Oldshue-Rushton contactors and Podbielniak centrifuges? What the heck is a Graesser Raining Bucket? The answer to all of these questions is "liquid extraction equipment."

Imagine heavy and light fluids flowing vertically past each other with only a very small density difference. Imagine rotating equipment that works better without rotating. Liquid extraction is different – very different. Do you know of any experts? The authors are looking for one.

(129a) Improved Filtration Analysis Applied to Filter Press Separations

Raymond Collins, Engineering and Process Sciences, Dow Chemical Co., Midland, MI

Improved Filtration Analysis Applied to Filter Press Separations

Raymond Collins, Engineering and Process Sciences, The Dow Chemical Co., Midland, MI

A non-traditional model describing the separation of solid particles from a liquid suspension by pressure-driven flow through a permeable barrier has been successfully applied to the design of several filter press applications. This flow model was derived by applying fundamental physical principles to the unsteady, mixed-phase system represented by cake filtration with a filter cloth. The analysis has proven suitable for operating modes employing both constant applied pressure and constant feed flow rate. The design procedure uses small-scale experiments to determine a governing permeability function with areproducible dependence on the collected volume of filtrate that is characteristic of the properties of the solids, liquid and filter cloth. The experimental method and various lab-scale test filters will be described, and the utility of this method will be demonstrated for a few common, large-scale applications in chemical processing. These include the removal of polymer fines and metal hydroxides from wastewater, the deliquoring of biomass from a concentrated sediment of fermentation broth, and the isolation and washing of micron-sized solids used to manufacture energy materials.

(129b) Uncoated and Polymer Coated Titanium Dioxide (TiO2) Nanoparticles: Differences in Deposition and Charge Characteristics

Biplab Mukherjee[1], Rajani Gourishetty[2], Wei Zhang[3], Termaine Powell[4], James Weaver[5], Richard Zepp[4] and Dermont Bouchard[4], (1)The Dow Chemical Company, Midland, MI, (2)Student Service Associate, EPA, Athens, GA, (3)Department of Crop and Soil Sciences, Michigan State University, East Lansing, MI, (4)Ecosystems Research Division, USEPA National Exposure Research Laboratory, Athens, GA, (5)National Risk Management Research Laboratory, EPA, Ada, OK

The advent of nanotechnology has opened up several potential avenues starting from the development of advanced manufacturing processes, revolutionary medical treatments, new consumer products, environmental applications, pollution control, etc. The backbone of this new leading-edge technology is the engineered nanomaterials (NMs), which are utilized both in the pure form (un-coated; U-NMs) and those for which the surfaces are modified, using polymers, surfactants, functional groups (coated; C-NMs), to attain certain specific properties in the final product. The colloidal behavior of U-NMs and C-NMs, even though when composed of the same core-material, can be drastically different and requires a deeper understanding for their optimal usage, in designing effective solid-liquid separation technologies, in minimizing exposure adversaries both to the human and the environment and for the development of reliable predictive models.

In this work, we have investigated the deposition and electro-kinetic behavior of both polymer coated and un-coated TiO_2 NMs, which directly impacts their separation and mobility characteristics. The deposition behavior of the NMs was investigated using deep-bed filters made using acid-washed sand columns and 1D convective-dispersive transport models. The effects of different forces (electrostatic, van-der Walls, and electrosteric--both osmotic and elastic) on the attachment & retention kinetics of NMs were analyzed. Predicted attachment efficiencies, using well-established dimensionless models, though matched well with the experimental values for U-NMs but failed for the C-NMs ($R^2<0.5$). However, an improvised model, which takes into account the effect of steric hinderance, improved the predicted attachment efficiency values for the C-NMs ($R^2>0.8$) and points towards the importance of other non-DLVO (Derjaguin and Landau, Verwey and Overbeek) forces for coated NMs.

(129c) Enhancing Filtration Rate of Phosphoric Acid by Polymers Application

Mazouz Hamid, Process and separation technologies, R&D OCP, EL JADIDA, Morocco

Phosphoric acid is primarily produced by what is known in the industry as the "wet process", the phosphate rock is fed into an attack tank or reactor and reacted (digested) with sulfuric acid (H2SO4). this reaction led to the production of phosphoric acid (H3PO4) and calcium sulfate dihydrate or gypsum(CaSO4·2H2O). Filtration is a key operation in the manufacturing process of phosphoric acid, the objective of this operation is to separate the phosphogypsum and phosphoric acid. My presentation will be subdevised on two main part, in the first part i will discuss the parameters that affect the filtration operation for the separation of phosphoric acid from phosphogypsum cake. In the second part, I will present a result of study for the filtration rate improvement and the reduction of P2O5 losses in the gypsum cake by using a suitable flocculant. The results of this study is based on the online mesurement of filtration rate and soluble content of P2O5 in gypsum cake, using bench scale of filtration unit. the study have been conducted in the case of two different process for phosphoric acid production Rhone Poulenc and Pryon.

(129d) Overview of Exergy Analysis for Distillation Operation

Kirk Liu, PE, Process Design and Development, UOP LLC - A Honeywell Company, Des Plaines, IL

Exergy analysis provides insights on opportunities for more thermodynamically efficient distillation operations. Typically, the use of this analysis has shown distillation column designs with new heat exchanger integration locations. This paper will provide a brief overview of the concepts for exergy analysis and simulation tools available to perform such an analysis. The paper will then end with a case review to demonstrate an implemented design with economic incentives from increased thermodynamic efficiency.

AUTHOR INDEX